Higgs Discovery

ALSO BY LISA RANDALL

Warped Passages

Knocking on Heaven's Door

Higgs Discovery

THE POWER OF EMPTY SPACE

Lisa Randall

An Imprint of HarperCollins*Publishers*

 For information address HarperCollins Publishers, 10 East 53rd Street, New York, NY 10022.

HarperCollins books may be purchased for educational, business, or sales promotional use. For information please e-mail the Special Markets Department at SPsales@harpercollins.com.

FIRST EDITION

Library of Congress Cataloging-in-Publication Data has been applied for.

ISBN 978-0-06-230047-8

13 14 15 16 17 OV/RRD 10 9 8 7 6 5 4 3 2 1

CONTENTS

Higgs Discovery

HIGGS DISCOVERY: THE POWER OF EMPTY SPACE

On July 4, 2012, along with many other people around the globe who were glued to their computers, I learned that a new particle had been discovered at the Large Hadron Collider (LHC) near Geneva. In what is now a well-publicized but nonetheless stunning turn of events, spokespeople from CMS and ATLAS, the two major LHC experiments, announced that a particle related to the Higgs mechanism, whereby elementary particles acquire their masses, had been found. I was flabbergasted. This was actually a discovery, not a mere hint or partial evidence. Enough data had been collected to meet the rigorous standards that particle physics experiments maintain for claiming a new particle's existence. The accumulation and analysis of sufficient evidence was all the more impressive because the date of the announcement had been fixed in advance to coincide with a major international physics conference occurring in Australia that same week. And what was more exciting still was that the particle looks a lot like a particle called the Higgs boson.

A Higgs boson is not just a new particle, but a new type of particle. The thrill in this particular discovery was that it was not simply a confirmation of definite expectations. Unlike many particle discoveries in my physics lifetime, for which we pretty much knew in advance what had to exist, no physicist could guarantee that a Higgs boson would be found in the energy range that the experiments currently cover—or even found at all. Most thought something like a Higgs boson should be present in nature, but we didn't know with certainty that its properties would permit experiments to find it this year. In fact, some physicists, Stephen Hawking among them, lost bets when it was found.

This discovery confirms that the Standard Model of particle physics is consistent. The Standard Model describes the most elementary components that are known in matter, such as quarks, leptons (like the electron), and the three nongravitational forces through which they interact—electromagnetism, the weak nuclear force, and the strong nuclear force. Most Standard Model particles have nonzero masses, which we know through many measurements. The Standard Model including those masses gives completely consistent predictions for all known particle phenomena at the level of precision of a fraction of a percent.

But the origin of those particle masses was not yet known. If particles had mass from the get-go, the theory would have been inconsistent and made nonsensical predictions such as probabilities of energetic particles interacting that were greater than one. Some new ingredient was required to allow for those masses. That new ingredient is the Higgs mechanism, and the particle that was found is very likely the Higgs boson that signals the mechanism's existence and tells how it is implemented. With improved statistics, which is to say with more information after the experiments run longer, we will learn more about what underlies the Higgs mechanism and hence the Standard Model.

Though a discovery was indeed announced, it was in fact made with some of the caution I had come to expect from particle physics announcements. Because the measurements had identified barely enough Higgs boson events to claim a discovery, they certainly didn't yet have enough data to measure all the newly discovered particle's properties and interactions accurately enough to assure that it is a single Higgs boson with precisely the properties such a particle is expected to have. A deviation from expectations could turn out to be even more interesting than something in perfect accord with predictions. It would be conclusive evidence for a new underlying physical theory beyond the simple model that implements the Higgs mechanism that current searches are based on. This is the sort of thing that keeps theorists like me on our toes as we try to find matter's underlying elements and their interactions. Precise measurements are ultimately what tell us how to move forward in our hypotheses. The Higgs boson is a very special particle indeed and we ultimately want to know as much as we can about it.

Whatever has been found—*the* Higgs boson, the particular implementation of the Higgs mechanism that seems simplest or something more elaborate—it is almost certainly something very new. The interest from the public and press has been very gratifying, indicating a thirst for knowledge and scientific advances that humanity to a large extent shares. After all, this discovery is part of the story of the universe's evolution as its initial symmetry was broken, particles acquired masses, atoms were formed, structure, and then us. News stories featured members of the public who were fascinated but weren't necessarily quite sure by what. Perhaps the ultimate recognition of the pervasiveness of Higgs boson awareness was the appearance of jokes and spoof news stories indicating the interest—but also some of the bewilderment.

So I'm writing this to respond to many of the questions I've been asked—to share what the discovery means and to explain a bit about where it takes us. Some of what I'll say is already in chapters from my previous books, *Warped Passages* and *Knocking on Heaven's Door*, two of which are appended. Those books didn't isolate the Higgs boson for extra special attention; rather they covered many topics, including information about the collider, the larger physics story for which this is the capstone, and the nature of science itself. They give the larger context of which this discovery is one—albeit a very important—part. But at least for the time being, the Higgs boson deserves its moment in the limelight. So in addition to those older chapters about the Higgs particle, this book offers a few new (and old) thoughts. It's an unbelievably exciting moment in physics and I'd like to share some of what occurred and what it means.

THE CHALLENGE OF DISCOVERY

I guess I was better off on July 4, 2012, than during the last Higgs report in December 2011. On the earlier date, I woke up before five in the morning to do an interview and to listen to talks from CERN, as I was in California and the time zone was not very congenial. At the time of the recent announcement, on the other hand, I found myself on a Greek island where I was taking an all too rare vacation. Although I had poor Internet connectivity and was isolated from my colleagues, at least I was only one time zone away when Joe Incandela, the spokesperson for CMS, first took the stage. Because my somewhat rustic apartment had no Internet, I first learned of the Higgs discovery while sitting in a balcony café—which happily for me opened at 10 A.M., the time of the talks.

In fact, I hadn't imagined when making my holiday plan that this would happen. I had known the Higgs evidence would increase, but I hadn't known that the engineers would have done such a heroic job in increasing the collision rate, and the experimenters an equally impressive improvement to their analysis methods, that would allow the speakers on July 4 to say with certainty (by physicists' standards) that a particle had been found. One other factor that contributed to the Higgs boson discovery was the decision to run at slightly higher energy—8 TeV rather than the 7 TeV of the previous year—which by itself increased Higgs production by about 30 percent. I was very grateful to the Internet for keeping everyone connected and to Twitter for providing an outlet for my excitement (and for sharing information once people caught on to what was happening and the connection diminished in quality).

Maybe to compensate for that disconnect, I spoke a few days later on a radio program on WNYC. In the pre-show discussion that one typically has before such a program, we reviewed the types of topics that might arise. Most were ones I was prepared for. But I was a little flustered when I was told I would be asked to compete with Dennis Overbye's delightful description of the Higgs boson discovery: "Like Omar Sharif materializing out of the shimmering desert as a man on a camel in 'Lawrence of Arabia,' the elusive boson has been coming slowly into view since last winter . . ." (*New York Times,* July 4, 2012).

I certainly wasn't going to think of something as magical in the half hour I had before the interview—especially as I love that movie. To further complicate the situation, I was on a rock-climbing cliff in Kalymnos, where I had to belay my partner up a climb that we had already set up (I didn't tell the producers that, since they would rightfully have worried about the connection).

So in anticipating my interview, I thought about the question. My partner suggested I say it was the Messiah whom physicists had been awaiting for fifty years, which I thought rather funny but not really helpful. I wanted to create an analogy that reproduced the physics better than as an actor or a deity representative.

What I came up with is not perfect but captures the lead-up to the actual discovery. I said it happened in the way you might find your friend in a crowded stadium full of shouting individuals, where everyone—including your friend—is making noise, but your friend, despite his distinctive voice, is a small peep in the noisy crowd. You would be hard-pressed at first to find him amid the huge din in the background. You might occasionally think you heard his voice, but then it would be drowned out or difficult to distinguish from that of others in the hordes of people.

But imagine now that you knew roughly where to look. You knew what section your friend was in and who he would be hanging out with. So you focus your attention in a particular region. When you hear the sound of his voice, you then begin to be increasingly confident that you have located him correctly. You might not know for sure right away but you might then begin to focus on an even more specific region of the stadium. Eventually you reach the right location where the voice is unmistakable and you know your friend has been found.

The Higgs discovery really did work like that. Higgs events are rare among the far more numerous ordinary particles that are produced. A Higgs boson is connected to elementary particle masses. That means that physicists can predict how it should interact (assuming of course that it exists), so you know its "voice." Alas, the interactions with the ingredients of a proton are in fact rather weak. Quarks and gluons experience the strong nuclear force, which is much more powerful than their interactions with a Higgs boson, so Higgs bosons will be produced only a small fraction of the time and get lost in the "crowd."

Out of the billions of particle collisions that occur every second, only rarely does a Higgs boson get produced. More often than not, collisions result in boring Standard Model collisions that we know to exist. Those collisions give us more detailed information about quarks and the strong nuclear force. But they muddy the waters for experimenters looking for a clear Higgs boson signal.

The only way to find the particle is to have a pretty good idea where to look so that experimenters can distinguish signal from the "din" of background. *Where* doesn't refer to a physical location like your friend's section in a stadium. Instead it refers to where in the data you expect Higgs evidence to lie—meaning which types of collider events are expected if a Higgs boson exists.

So as with your friend in the stadium, the Higgs boson was

initially lost in the background data. Experimenters subsequently looked through trillions of events so that they could begin to see evidence for a bit of a deviation that could signal something special. This was something like the December announcement, where evidence for a particle was presented (at only 3 sigma level, 3 standard deviations, which is a statistical term meaning the probability is less than one in a few hundred that the result isn't a signal). For those interested in the precise criteria, the level 3 sigma is thought to be too small for discovery, mostly based on experience. Sometimes experimenters don't yet understand all the details of their system or prediction, fluctuations do happen, and everyone knows that if they wait, the answer will sort itself out. That is why experiments require a signal more than 5 sigma, which means that the odds are less than one in 1.7 million that it is simply background noise arising from known familiar particles.

But as time went by and more data accumulated, physicists zoned in on the region that looked a little different and sorted through at least twice as much data. With enough data and enough understanding of the properties expected for a Higgs boson, a clear signal emerged. That signal was what was revealed on July 4 and is almost certainly connected to the Higgs mechanism by which particles acquire their masses.

THE HIGGS MECHANISM, THE HIGGS FIELD, AND THE HIGGS BOSON

In order to fully appreciate this discovery and what there remains to explore, it helps to know a bit of particle physics, including some deep and subtle underlying concepts. Of critical importance is the distinction between the Higgs mechanism, the Higgs field that is involved in the mechanism, and the Higgs boson particle—which is what an experiment can actually find. Even without experimental proof, physicists were fairly confident about the mechanism, since it was the only consistent way to give elementary particles their masses, as the attached chapters explain in further detail than I do here.

But despite the theoretical consistency of the idea and the failure of any other idea to explain masses, physicists all wanted experimental proof. The experimental results from the LHC have now rather firmly established the relevance of the Higgs mechanism and the Higgs field on which it relies. They have also established the existence of a new particle related to the mechanism—but *the* Higgs boson is part of a very particular implementation, which only further data will definitively confirm or rule out. That is why my title, *Higgs Discovery*, is deliberately ambiguous.

The Higgs mechanism is responsible for elementary particle masses, such as the mass of the electron. Mass is what provides resistance when a force is applied. If particles have no mass, they travel at the speed of light. A particle's mass tells us how it responds to forces and how it travels through space.[1]

Without its mass, the electron wouldn't bind into an atom, and just about everything else you take for granted about the world wouldn't work either. Such an elementary particle mass, that of the electron in this case, relies on the existence of what particle physicists call a field—a quantity that exists throughout space but doesn't necessarily involve any actual particles. Admittedly, the concept of a field is a bit esoteric and confusing, especially as the word *field* outside of physics conjures images of cows grazing, which became clearer to me when I read the word *champs* in French physics textbooks.

But really we encounter fields (in the physics sense) in many contexts. A magnetic field is perhaps the most familiar one. When you hold a magnet close to your refrigerator door, you feel a force. There is "nothing" between the magnet and the refrigerator, which is to say there is no actual matter there. But there is a field and that field is responsible for the tug you feel that attracts the magnet to the shiny white surface.

That field is of course local. You move a little away from your kitchen and the influence of the magnet is too negligible to feel. The Higgs field, on the other hand, is everywhere. It is spread throughout the universe. The field isn't made up of actual particles. In a sense it involves something like a type of charge spread everywhere through empty space. Particles that experience the weak force (which is to say leptons like the electron, quarks, the weak gauge bosons that communicate the weak nuclear force, and the Higgs boson itself, as we will soon see) interact with that "Higgs charge" and thereby acquire mass. In the presence of the Higgs field, particles have masses. The heaviest particles interact with the field the most, and the lightest the least.

The Higgs boson, on the other hand, is an actual particle—a fundamental object that has definite mass and interactions. Although not a field itself, the particle is indeed associated with the Higgs field. Essentially when you jiggle the Higgs field—add a bit of energy—you can create an actual particle. A single field both permeates the vacuum—empty space—with a nonzero constant value everywhere, and is also responsible for particle creation.

From the point of view of particle masses acquired through the Higgs mechanism, the Higgs boson is an appendage that comes along for the ride. The Higgs boson itself is not essential to particle masses. Clearly masses didn't just come into being on July 4, 2012, with the announcement of discovery. The Higgs field—not the Higgs boson—gives masses.

Yet the Higgs boson is the telltale sign that the theory of the Higgs mechanism is realized in nature. With enough energy, a Higgs field can create these particles, and they have properties specific to the role of the Higgs field in allowing for particle masses. On July 4, we learned that the Large Hadron Collider at CERN had finally made enough of these Higgs bosons to give physicists a clear signal that couldn't be mimicked by Standard Model particles we already knew to exist. In other words, the signal could make sense only with a new particle present.

The Higgs boson—or a Higgs boson surrogate whose properties deviate from that of the simplest implementation of the Higgs mechanism—is also essential to our learning how the Higgs mechanism actually came about. Future data are guaranteed to offer more details about the particle's properties so that we can definitively establish whether it is part of a simple sector of particles implementing the Higgs mechanism, or an even richer one, involving more Higgs-like particles or additional structure involving new forces and interactions.

It's interesting to reflect on the implication of this discovery for our understanding of "empty" space. When we describe space as empty, we generally mean that no actual matter is present. Matter is stuff that clumps together under the force of gravity to form structure. The Higgs field is not like that. It takes a nonzero value but remains uniformly spread everywhere.

The Higgs field is not composed of actual matter. In a sense the Higgs field carries a type of charge and furthermore allows that charge to appear and disappear into it. We don't directly feel that charge (associated with the weak nuclear force) because the force associated with it has such short range. It does, however, allow for nuclear beta decay, for example, by which a neutron can decay into a proton, an electron, and a neutral particle called a neutrino.[2]

Particles have interactions directly with the Higgs field that don't involve the Standard Model forces at all. In a sense, the Higgs boson itself communicates a kind of force, but one very different from those we know about involving only a fixed unit of charge. Particles' interactions with the Higgs field allow for a wide range of masses because each particle has its own individual interaction with the field.

You can ask whether the Higgs field carries energy too. We don't know the answer, since all that can be measured is the gravitational influence of the net energy of all the fields in the universe. This energy takes a nonzero value known as the dark energy, which is a rich and fascinating subject in its own right. The dark energy is also associated with empty space—it is the energy empty space possesses. Einstein taught us that any source of energy has consequences for gravity, including the absolute energy of an empty universe. The energy carried by empty space has measurable consequences, such as the acceleration of the expansion of the universe. The fact is, empty space is not truly empty. It can have energy and charge. It just doesn't have matter.

Even though we don't know the energy carried by the Higgs field, we do know that the energy depends on the value the Higgs field takes. The energy is lowest when the Higgs field is nonzero. This is discussed more completely in an attached chapter, but the field is a bit like a pencil standing on end. The zero value is analogous to the pencil standing upright, whereas the nonzero value, which carries lower energy, is analogous to the fallen pencil lying on its side. The Higgs field too doesn't want to be at the symmetric point with zero value but prefers to "fall" in some direction where it breaks a symmetry.[3] When it does so, it takes on a nonzero value that is ultimately responsible for elementary particle masses. Admittedly this is all a bit abstract in nature, since it doesn't involve matter we can throw around and touch. But one of the beautiful aspects of the Higgs mechanism is that it tells us about the richness of empty space.

Before ending this section, I'll answer one more interesting question that I've been asked. Where does the mass of the Higgs boson itself come from? The answer is that the Higgs boson interacts with the Higgs field. So just as with other elementary particles, the Higgs field accounts for the Higgs boson's mass.

You can think of a physical interaction as two particles entering a collision, and then two particles emerging, most likely in another direction. That is what happens when an electron scatters off another electron, for example. In the case of the Higgs boson, the same interaction that permits two Higgs bosons to collide and merge also allows a single Higgs boson to interact with a background field—the field existing in empty space. This is key to understanding one of the questions physicists ask as we move forward: What does the mass tell us about the Higgs' interactions with itself? This self-interaction, as it is known, might be determined by physics that goes beyond the Standard Model and is one more important clue—in addition to the different Higgs boson decay rates—of the nature of the particle and what lies beyond.

HIGGS BOSON DECAYS

A key property of the Higgs boson—one that is essential to understanding how it is found—is that it is extremely unstable. It lasts only a fraction of a second before turning into Standard Model particles such as quarks and leptons. *Decay* means that the Higgs boson itself ceases to exist and instead decay products (familiar Standard Model particles) are created and carry away its initial energy and momentum.

That means that when experimenters search for a Higgs boson, they don't look for the particle itself but for the particles into which it decays. By adding up the charge, energy, and momentum of those final state particles, they can determine if their origin was a particle with definite charge (zero in this case) and a particular mass.

If a real particle decays, there is a real mass involved. So when you plot the number of events you have versus putative mass, you would find a bump—an excess of events that is centered at the actual mass of the physical particle. That is what happened with the Higgs discovery, which is why you might have heard experimenters discussing their discovery in a language using the word *bump*.

The bump isn't just a line, however, which is what would

happen in a world where all measurements are perfect. Because they are not, the measurements are centered on the physical value but there is a smooth drop-off away from the exact Higgs boson mass—a drop-off characteristic of slight mismeasurements. With increased data, the measurement improves and the signal becomes a narrower peak.

In fact, even if measurements were perfect, there would still be a spread of values, but a much smaller one—because of quantum mechanics. The Higgs boson decays; it doesn't last forever. The quantum mechanical uncertainty relation permits the Higgs boson mass to look wrong for a short amount of time (less than its lifetime). Those Higgs bosons with slightly wrong masses get recorded to generate what is known as the Higgs width. That width is in fact a measure of the Higgs boson lifetime. Unfortunately, at the LHC, it is too small to measure, since the larger experimental uncertainties give more uncertainty to the mass that is measured.

Actually, as if experimenters didn't have enough on their plates trying to find the Higgs needle in a haystack, there is one more confounding problem that makes measurements even more challenging. This problem is known as pileup. The LHC is a remarkable machine for at least two reasons.[4] One is that it has higher energy than any accelerator that existed before. The other is that it has a remarkably high intensity. That is, the rate of events is enormous. In fact, it is so enormous that more than two protons collide when a bunch of protons within the beam collide.

At full intensity, the number of protons is enhanced to the point where about thirty collisions occur at virtually the same time. Most are uninteresting and don't confuse the data too much. But when doing a precision study such as a Higgs boson decay, you can't be too careful, and the experimenters employed a number of clever techniques to identify which particles are really part of the collision of interest. That involves identifying the right decay products that came from the Higgs boson decay. The next section is a more nitty-gritty one that tells what those might be.

DECAY MODES

A Higgs boson decays, converting into other particles that carry away the particle's initial energy, momentum, and charge. Detectors such as those in ATLAS and CMS measure all those properties of the decay products which fly away from the region where the protons initially collided. Experimenters have to add up the energy and momentum of all the particles emanating from the collision region where the Higgs boson is produced and decays to figure out the properties of the particle that was momentarily in existence there.

The connection between the interactions of the Higgs boson particle and Higgs boson field allow physicists to predict the Higgs boson's interactions based on particle masses. Those interactions are important to studying the Higgs boson because they are what permit the Higgs boson to decay.

The particles into which the Higgs boson decays also have to have a sufficiently large interaction with the Higgs boson that such a process can occur. If a Higgs boson didn't interact with a particle, it couldn't decay into it. The Higgs boson is a very special particle whose interactions are connected to particle masses: heavier particles interact more and lighter ones interact less. So a

Higgs boson decays the most into the heaviest particle that is not so heavy that energy won't be conserved in the decay. The bigger the mass, the bigger the interaction.

But if the mass of the particle is too big, there won't be enough energy to make it. The particles into which the Higgs boson decays must be sufficiently light to allow energy to be conserved in the decay—along with momentum and charge, which the decay products all carry away.

The bottom quark is the heaviest particle for which twice the mass is still less than the measured mass of the Higgs boson—125 GeV—so that a bottom quark and its antiparticle, the bottom antiquark, can be produced. Because it is the heaviest particle for which the decay can occur, most Higgs boson decays are into bottom quarks and antiquarks.

To deconstruct that paragraph: GeV is a funny unit of mass that particle physicists use. It is actually a measure of energy. GeV means giga electron volts, which is a billion electron volts. We also sometimes speak of TeV, which is the equivalent of 1,000 GeV. The collider currently runs with 8 TeV of energy. Energy can be used as a measure of mass too. Einstein's famous formula $E = mc^2$ tells us that energy (E) and mass (m) can be used interchangeably, since c, the speed of light, is constant.

The bottom quark is a type of elementary particle. You might have heard of the up and down quarks that sit inside a proton and a neutron. That is to say, the protons and neutrons sitting inside the nucleus of an atom are not elementary but are in turn composed of more fundamental particles called quarks. Those quarks are held together by the strong nuclear force—yes, that same force I referred to earlier that often produces Standard Model particles when protons collide. They are the two lightest quarks. One has positive charge (the up quark with +2/3 charge), and one has negative (the down quark with charge -1/3). Two up quarks and a down quark together give the proton its charge of +1.

But those aren't the only quarks. One of the biggest mysteries in particle physics is that for every familiar type of particle (by familiar I mean ones that exist on Earth today), there are heavier versions that have bigger mass, are unstable, and are created on Earth only in accelerators and, for some, in cosmic rays.

The bottom quark is one of those heavier quarks. Like the down quark, it is negatively charged and in fact is the heaviest of the three quarks that carry that same charge (the other is the strange quark). The reason the Higgs boson decays primarily into the bottom quark is that it has this relatively heavy mass and therefore interacts more with a Higgs boson, whose interactions are determined by particle masses.

But unlike a bottom quark, a Higgs boson has zero charge. If charge is to be preserved, the Higgs boson can only decay in a way for which the net charge of the decay produced—the sum of the charges of all the particles into which it decays—is zero. An antiquark carries precisely the opposite charges of a quark. And a bottom antiquark carries precisely the opposite charge of a bottom quark. The charges of the bottom quark and its antiquark add to zero, which is the charge of the Higgs boson that decayed into them.

That is good because in addition to electric charge, a bottom quark carries another type of charge related to the strong force. A bottom quark is not neutral under either the electric force or the strong nuclear force. However, a bottom quark and a bottom antiquark together are neutral. A bottom quark and antiquark together can carry no net charge whatsoever. That is exactly the property of a Higgs boson—it carries no net charge. A Higgs boson does interact directly with quarks and other elementary particles that have masses. But those interactions are not merely through the three known Standard Model forces (electromagnetism, the weak nuclear force, and the strong nuclear force; I'm leaving out gravity since it's so extraordinarily weak). Direct interactions related to those that are responsible for masses always occur.

So a Higgs boson can decay into a bottom quark and a bottom antiquark without violating any known conservation law. And it does just that. Of course, the Higgs boson interacts with heavier particles too—the weak gauge bosons and top quark in particular. But a 125 GeV Higgs boson isn't heavy enough to decay into those particles. The sum of a top quark and antiquark mass is far in excess of 125 GeV. The sum of two charged weak gauge boson masses is about 160 GeV, which is again too heavy for a Higgs boson to decay directly into those particles.

So the dominant Higgs boson decay is into bottom quarks. But here's the rub. At the LHC, bottom quarks are tough to identify and distinguish from the Standard Model "din" (for physicists, known as background). They are quarks, and so many quarks get produced at the LHC that you need to look at some special production mode in order to isolate the Higgs decay signal.

When the Higgs boson announcement was made, the two modes that dominated the signal were decays into photons and decays through weak gauge bosons—not the more frequent decays into bottom quarks.

I realize this is a lot of information, but if you're a little perplexed at this point, you should be. I said the Higgs boson interactions correspond to elementary particle masses so that it should interact more with heavier particles. The photon has zero mass, which makes it seem as if the Higgs boson shouldn't interact with photons at all. On top of that, I just said the Higgs boson is too light to decay into weak gauge bosons. So what am I talking about?

I'm talking about quantum mechanical effects. The most visible decays of the Higgs boson to date occur because of processes that are allowed only when we take quantum mechanics into account. The Higgs boson can in fact decay into two photons, but only because quantum mechanics permits it. When the CMS and ATLAS experiments were designed, a strong consideration was being able to measure energy and direction of photons as well as possible to allow for a precision Higgs boson discovery.

Quantum mechanics allows a Higgs to decay into two photons because it allows for virtual particles—particles that don't exist forever and in fact don't even have the correct mass to be the actual particle that can survive in the real world. Virtual particles are a big deal and one of the many nonintuitive properties that quantum mechanics permits. They are particles that briefly come into existence, can have further interactions, and then must disappear.

Now, even though a Higgs boson wouldn't interact with photons according to classical physical laws (those that don't take quantum mechanics into account), it will interact with them according to quantum mechanical laws. What happens is a Higgs boson turns into particles too heavy to be made in nature, such as weak gauge bosons and top quarks—the heaviest of the Standard Model quarks. The heavy particles disappear, annihilating each other, but in doing so emit photons. Even though the interaction between a Higgs and two photons is suppressed (all interactions permitted only through quantum mechanical processes are), the overall rate for decay into photons is small but not negligible. The Higgs boson decays into photons about a fifth of a percent of the time. And this is sufficiently often to generate a clear photon signal once enough Higgs bosons have been produced.

Let's now return to the decays involving weak gauge bosons—the other decay that was important to discovery. Those in fact happen a little differently. I wish this were all more straightforward to explain, but I'm giving you the real story here. Weak gauge bosons also get produced as virtual particles—those that have the wrong mass to be the actual physical particle that survives in the real world. But they can have almost the right mass. After all, half of 125 GeV—the amount of energy each weak gauge boson has at its disposal—isn't that far from 80 GeV—the mass of a charged W boson, or from 91 GeV—the mass of a neutral Z boson. It's sufficiently far away that two real gauge bosons cannot be produced. But it's sufficiently close that the rate of decay through two weak gauge bosons, after which the weak gauge bosons themselves decay, is in between that of a classical and a quantum mechanical process. So for this particular mass of a Higgs, decays through weak gauge bosons occur reasonably often as well.

Experimenters look for these decays by searching for the decays of two gauge bosons. These are not just any two gauge bosons, however. They are gauge bosons that emerged from the decay of a Higgs. This means that when you add up the energy and momentum of the decay products and put it all together, it will reproduce the mass of a Higgs boson. So these events are distinguished from other events by contributing to a Higgs boson bump when you plot the masses that experimenters put together from the decays.

THE DEVELOPMENT

So now that we know a bit more about what experimenters actually look for, let's return to how the discovery played out over the course of the past seven months.

The LHC has now been running for about two and a half years. One of the chief search targets all along has certainly been the Higgs boson. The interesting thing about the Higgs boson was that combined experimental and theoretical considerations told us to expect it to be relatively light, which is to say well within the kinematic reach of an LHC that was running at only half energy. We didn't know for sure this would prove correct, but it was promising.

However, because the particles that create the Higgs boson are light (and therefore have small interactions with it), and because the Higgs boson can be best observed in modes of decay that happen only infrequently, experimenters needed many collisions before they could see a signal that sufficiently dominated over background. So everyone had to wait as data were collected.

Of course it was worth it. The first real hint that something was afoot (at least to those of us not on the experiments) came in December 2011. Even in my sleep-deprived state many time zones away in California, it was clear that something exciting was occurring. There was evidence that a particle was decaying to two photons at a rate that exceeded the Standard Model expectation in the absence of a Higgs boson. ATLAS especially had something that looked like a signal, while CMS had data that certainly didn't rule out that possibility but on its own might not have looked as strong. The CMS data improved over the next few months, however, so both experiments had evidence of a signal of something new—just not strong enough to claim discovery.

The reason physicists require a big signal to claim discovery is that the search requires evidence of an excess of events over background—the *bump* I referred to earlier. But no one can predict exactly what happens in any particular collisions. The predictions apply to the average. With only a few collisions, the data might contain a spurious signal—a fluctuation that exceeds background expectations and accidentally mimics a Higgs boson. Only with enough data—enough collisions—will the probability that you are seeing just a fluctuation become sufficiently unlikely that you can really believe you have discovered a new particle.

Personally, I was gratified by the December result. It was the best we could have hoped for at the time unless our predictions were wildly wrong. There simply hadn't yet been enough collisions to make a true discovery so it is what I expected when I heard hints that there could be a signal. In fact, I'm on record as saying this since I answered Dennis Overbye's questions for the *New York Times* and did a radio interview immediately before the announcement—a bit of a risk since as a theorist who isn't a member of any experiment, I didn't have much more information than anyone else until the result then was made public.

The subsequent months were very exciting for physicists. Theorists like me had to take into account limits set by the LHC and the possible existence of a 125 GeV Higgs boson, and I and many others developed models that did so. In January I visited CERN, the major physics center near Geneva that houses the LHC, and had valuable conversations with physicists there. Experimenters shared their insights and were eager to learn more about how to search for models that I was working on at the time.

In March I had yet more contact with the experimental community at a major international conference named Moriond, that delightfully takes place at an Italian ski resort (which suffered, I'm afraid, from prematurely warm weather). People there were still discussing "faster than light neutrinos" measured by the OPERA experiment, which are definitely excluded now. But they had only speculation and some modest measurement improvements to share about the Higgs boson. The physicist Greg Landsberg expressed his frustration at needing to talk about Higgs results after drinking "phantom of the opera" grappa. There really was no discovery to report, and the updates were not overly exciting in March.

In fact, on asking around, I found that virtually no one thought a discovery would be announced by the time it was. Discovery arrived more swiftly because engineers led by Steve Myers managed to crank up the intensity of the machine and experimenters significantly advanced their analysis techniques.

As for whether people believed the Higgs result of December was real, the majority expected it to be confirmed. However, you might be surprised to learn that many theorists had hoped the signal would prove incorrect because of the deep consequences it would have for the underlying theoretical construct. If there were no Higgs boson, consistency would have required something even more surprising and interesting.

I have to say, given my dealings with the public—and with experimenters—I found it hard to share in that hope, interesting as it would be. Experimenters deserved to find something. They had worked long and hard—and waited a great deal too. Setting limits is important, but a discovery is something else altogether. Finally they have a particle to measure more about and understand.

And, I must confess, as challenging as it is to explain why a Higgs boson discovery is interesting, explaining why not finding it would be more interesting was a task I was glad to avoid.

One of the funny things about the Moriond conference, however, was that even though so much attention was focused on the Higgs boson, most talk titles used the words "scalar particle" or "scalar boson" rather than "Higgs boson." I later found out that this was because François Englert, who was one of the original six physicists—Peter Higgs, Robert Brout, and François Englert together, and the collaboration of Gerald Guralnik, C. R. Hagen, and Tom Kibble—who developed the so-called Higgs mechanism, was in attendance as well. I will not enter into a discussion of priority claims, but the mechanism was the capstone to an edifice that has been in the making since Marie Curie first discovered radioactivity, showing that more happened at the subatomic level than anyone had imagined, and culminating—as of today—in the Standard Model and the Higgs mechanism.

Englert gave a wonderful talk about the particle and the

mechanism, and I was surprised by how engaged I found myself, even though I was familiar with the material. One of the true delights at Moriond was getting to know François Englert a bit better. Peter Higgs is an interesting character too, but I don't know him myself. When I visited Bristol for the Ideas Festival there, I learned that Peter Higgs's interest in physics was sparked in part in Bristol, where he attended the same school as Paul Dirac (the physicist who developed the idea of antimatter).

Higgs had originally planned to study engineering—as did Dirac for that matter. Having then decided to study particle physics, he was initially dissuaded and studied molecular physics before happily returning to the type of physics that truly interested him and making his major breakthrough almost fifty years ago in Edinburgh.

The Belgian François Englert had perhaps an even more surprising story. He too had studied engineering and had been sent to England "to work on cables," as he described it. Apparently his major accomplishment there was to incite a strike, for which he was rewarded with pay on condition that he leave right away. Having had little interest in cables in the first place, Englert told me this was an offer he could not refuse. He returned home and started to work on semiconductors, which was closer to the physics he would ultimately study, but he was working in a laboratory and not doing the type of theoretical work that would continue to hold his interest.

Englert's next step was the Belgian army. Fortunately for him, it had a good library and he managed to find an adviser in condensed matter physics and get a degree while doing his required service. When he moved to Cornell for a postdoctoral fellowship, he had the good fortune to work with Robert Brout, who would become his major collaborator. The collaboration was so successful that Cornell offered Englert a job, which he refused in favor of returning to Brussels, where he did not yet have employment. And more remarkably still, Brout—though tenured at Brussels and not yet employed at Cornell either—decided to join him. In any case, the story continued remarkably satisfactorily. They both got jobs, and they went on to do work for which Englert is likely to win a Nobel Prize (sadly, Brout recently passed away). They played a significant—and daring—role in Belgium, where the new ways of thinking about elementary particle physics had not yet caught on.

I also had the satisfaction of introducing François to a young French professor who worked on the Higgs boson search and watching Englert's interest and excitement at the then current experimental situation as they discussed the physics into the night. Clearly the moment of truth was approaching for Higgs, Englert, and the other potential laureates involved in the Higgs theory development, as well as for the rest of us.

So what happened between March and July? Clearly, quite a bit. The engineers and experimenters were hard at work. Theorists were too, but we were not privy to the daily updates in data, since experimenters don't want false rumors leaking out. So I personally did not know much more about what was to come. I knew the Higgs result would improve because someone had told me about a smile a lead experimentalist had when viewing the data (yes, that is the kind of tea leaf reading we are sometimes reduced to), but I didn't know a discovery would be announced.

In fact, I was not alone. At a European conference I attended a week after the announcement, I had the opportunity to talk to Rolf Heuer, the director general of CERN, and Steve Myers, the chief engineer. Both told me of the uncertainty—and excitement—of the days and weeks preceding the announcement. Every day the results were different—though converging. When the CERN seminars were scheduled only two weeks before July 4, no one was yet certain of the results. In fact, even up until the last week and the last few days, the numbers were still fluctuating. Claiming a discovery is a very big deal, one that everyone involved took very seriously. It was only in the last days that it became clear that word could be used.

Of course, by July, members of the experiments knew what was happening. At CERN, many others certainly did know something was afoot by the time of the announcement. Students and others who were not sufficiently elevated in the CERN hierarchy even camped out the night before to guarantee themselves a seat in the auditorium—which turned out to have been worth it when, in the morning, hundreds of people were turned away.

Joe Incandela, the spokesperson for CMS, gave the first talk. He began by discussing the modes with the strongest signal, the decay into two photons and the decay into neutral weak gauge boson. After calmly discussing the results, he put a slide up saying the two modes together gave the 5 sigma signal that had come to be the physicists' benchmark heralding discovery. It was extraordinary. The CERN audience broke into applause when those words were spoken. In my outdoor café, I could do no more than tweet to vent my excitement.

The ATLAS talk followed and their results were just as exciting—even if the Comic Sans font designed for children that was used in the presentation was the source of great amusement on the Internet and among my tech-savvy friends. After Joe Incandela's talk, Fabiola Gianotti, the spokesperson of ATLAS, spoke about their 5 sigma and got applause too. In fact, she had to remind the audience that her talk wasn't yet over so that she could finish.

The remarkable conclusion of the seminars was that both CMS and ATLAS had made a discovery and it was very likely connected to the Higgs mechanism. Rolf Heuer happily summarized the situation after the two talks. We don't yet know for sure whether it is the simple sort of Higgs boson first proposed or something more complicated that plays the same role, but we do know that a particle connected to the Higgs mechanism has been found.

In fact, both experiments have been lucky (or there is something very important yet to be understood) in that the signal in two photons seems to fluctuate upward from what is expected in the Standard Model (meaning there was a slightly larger bump than was expected). For CMS, the photon channel is high but the neutral gauge boson signal is low. So are signals in a couple of other types of decays that ATLAS hasn't even checked yet. In ATLAS, not only did the photon data exceed expectations, they were lucky with the neutral gauge boson decay mode too.

Gianotti pointed out during her talk that we are very fortunate with the Higgs boson mass. There are in fact five different decay modes that occur often enough to be studied in detail. Had the Higgs boson heavier, decays to weak gauge bosons or top quarks would have overwhelmingly dominated. Had the mass been any lighter, the decays through virtual weak gauge bosons would have constituted an insignificant fraction of the total decays. But for this particular mass, essentially all measurable decay modes are rare. No single one dominates the data. Nonetheless, all occur with enough strength that future LHC running will permit detailed investigation. This is critically important to establishing whether the discovery was of a single particle called the Higgs boson or was part of a bigger story in which Higgs boson properties aren't exactly those predicted in the simplest model.

In fact, I'll take this opportunity to make a brief digression about particle accelerators. Before the LHC was constructed in Europe, Americans had built the Tevatron, a smaller accelerator that collided protons and antiprotons at one-quarter the current energy of the LHC (and one-seventh the LHC's ultimate energy). The energy and collision rate were not sufficient to discover a Higgs boson at the level of rigorous standards we particle physicists require. But they certainly could have measured Higgs boson decays. Furthermore, this machine was a proton-antiproton collider, which made a particular Higgs production mode much more important. So unlike the LHC, the best Higgs signal was in bottom quarks.

So although today we celebrate the LHC's discovery, my hat is off to the Tevatron experimenters too. They did in fact measure an excess of events involving bottom quarks that seemed to be centered at the right mass to correspond to the Higgs boson, but much less than required for a discovery. Even so, Heuer remarked to me the reassurance that came from the supporting evidence the Tevatron provided.

Although the Tevatron has now ceased to operate, the LHC will provide lots more information about the different Higgs boson decay modes, including decays into bottom quarks, photons, weak gauge bosons, and tau leptons (heavier versions of the electron). Further measurements of the different decay modes should determine whether or not the slight deviations from Standard Model expectations are real reflections of the physical world or mere statistical fluctuations in data that will disappear over time. This will require more collisions.

CERN's plan was to shut down the LHC for a year and a half at the end of the year, during which time the machine will be refurbished and upgraded. When the machine turns back on, it will have considerably higher energy. This higher energy will give the LHC a huge boost in the search for new particles and new physical theories.

The good news that Rolf Heuer shared within a week of the Higgs announcement is that CERN will keep running the LHC a few months beyond the previously planned shutdown. With this increased running time, experimenters should be able to collect enough data to make the necessary determinations on the nature of what has been found. A few months might not sound like much, but the increase in the collider's intensity means that later data carry more weight than earlier data. With enough collisions, the properties of the Higgs boson will be much better known, and that is good news for those of us who want to know what direction to head in during those two years while the collider is dormant. Either way, a large community of physicists and others look forward to the result.

So new data might soon tell us if the particle that was found is *the* Higgs boson or if it provides hints of something beyond the Standard Model. It might also tell us more directly about other new particles and interactions, though realistically that might require at least the higher energy (13–14 TeV) that the LHC will have after the two-year shutdown and upgrade.

Whatever is found, physics will move forward. We will either know that the simple Higgs boson has been found. Or we'll have evidence of physics beyond the Standard Model that will point the way forward. These detailed studies are very well motivated. The Higgs boson discovery is more likely to be the beginning of the story than an end.

A FUNDAMENTAL SCALAR?

Now that a Higgs particle has been found, of course the big question is where to go from here. Yes we can take a few moments to celebrate and drink champagne, as LHC experimenters tend to do. But ultimately physicists want to put together a bigger picture.

I've discussed the existence of a Higgs field, which is a quantity that permeates the vacuum and has a nonzero value that is the root of elementary particle masses. A nonzero value for a field is indeed something special. If a field carries charge, it means that charge can disappear into the vacuum, so the charge won't be conserved. If the field changes under rotations, much as an arrow pointing in a particular direction would do, the vacuum wouldn't preserve rotational symmetry. And if the particle that is created by the field has nonzero spin, the rotational symmetry (and Lorentz symmetry, Einstein's extension of rotational symmetry that includes time) would be broken as well.

This means that the particle created by the field also has to be special. It has to have zero spin, which is to say it must be a scalar particle, which is a particle with spin 0. Spin is the property of a particle that tells us how we can expect it to behave under rotation. Particles like the photon have spin 1. Particles like the electron or a quark have spin ½.

For particle physicists, spin 0 is equivalent to saying that the Higgs particle (created by an associated field) doesn't change under spacetime symmetries such as rotations. An electric field turns on only when actual charged matter is present. Otherwise, rotational symmetry would be broken. The Higgs field with zero spin can turn on in the absence of any particles whatsoever, since no symmetry (such as rotational symmetry) is broken when it does. The Higgs field is the same in all directions.

The reason this is so potentially interesting is that up until now, no one has discovered an elementary scalar particle. Scalar particles do exist, but the ones people have observed so far are composites made up of more fundamental particles such as a quark and an antiquark.

The Higgs boson has the potential to be the first elementary scalar particle ever found. This means that the Higgs boson in some sense is a truly new form of particle. You can even think of particles that interact with it experiencing a new type of force, one distinct from the four known forces.

But there is a mysterious side to a fundamental scalar. According to calculations based on quantum field theory, which combines quantum mechanics and special relativity, fundamental scalars should be extremely heavy—sixteen orders of magnitude heavier than the boson that has been measured. Such calculations indicate that the Higgs boson should be enormously heavier than we know it to be.

Even before the LHC measurement, physicists knew the mass should not be very different from the masses of the weak gauge bosons—near an energy associated with a symmetry breaking, which is about 250 GeV. I want to be clear that this is an approximate criterion. We don't necessarily require precisely a 250 GeV mass; 125 GeV is just fine. But we don't want the mass prediction to be ten million trillion GeV. Nonetheless, without additional underlying physics, a light fundamental scalar is an enormous fudge, or what we call "fine-tuning."

This is the hierarchy problem of particle physics described in an appended chapter (and discussed further too in *Warped Passages*). The answer to this conundrum is likely to involve some deep new understanding of nature, which could be a new symmetry of space and time, such as the theory known as supersymmetry, or even an extension of space itself, such as a warped extra dimension.

In any case, even if *the* Higgs boson exists, it is most likely part of a larger sector of new particles. That would be a big story beyond the Higgs that we hope to learn more about next from the LHC.

NOW THAT WE'VE FOUND HIGGS, WHAT DO WE DO WITH IT?

We will all learn more as the LHC continues to run. With further data, we'll learn if indeed *the* Higgs boson has been discovered. We will also have further constraints on new physical theories as searches continue if they don't find anything, and of course if they do find something else new, it will point the way forward.

Really, although most people hesitate to claim the particle that was discovered is *the* Higgs boson—a particle in a specific model that implements the Higgs mechanism responsible for elementary particle masses—the discovery was made because the properties look an awful lot like those we would expect from this particular particle. At the level of accuracy of the measurements, the data conform to Higgs boson expectations, giving few hints so far as to what lies beyond the Standard Model. Even if it turns out to be part of a more complicated model than the one that gives rise to a single Higgs boson and nothing else, whatever is found is part of a theory associated with the Higgs mechanism yielding elementary particle masses.

However, until all the different decays are measured with better precision, it still isn't absolutely clear whether a simple Higgs boson of the sort that seems to have been found was responsible for the Higgs mechanism or something more complex. But even what we know so far severely constrains the possibilities that theorists like myself can now think about.

On top of that, the Higgs boson, even if it is *the* Higgs boson, is almost certainly not the only particle yet to be discovered. The LHC was not designed simply to look for a single particle. It is searching for yet richer elements that can underlie the Standard Model. Knowing that the Higgs boson is part of the sector associated with the Higgs mechanism certainly addresses one mystery—how elementary particles acquire their mass—but still leaves open the question of why those masses are what they are.

Forging ahead, there are several agendas both for experimenters and for theorists. Experimenters will have to measure this particle's properties with much better precision. That will allow us to determine with greater certainty what in fact has been found. They will also be searching for extensions of the Standard Model that go far beyond this one additional particle. Any extension will almost certainly involve a sector with many additional particles and hopefully many additional experimental signals.

Theorists on the other hand will continue to pursue the hierarchy problem. But we will also puzzle over the particular value of the mass that now seems to be established. Some of my recent research involved trying to reconcile what the LHC has measured with proposed theories that have already been suggested. New data might quickly obviate any such speculations. But they give a flavor of the kind of work we do. And of course if we're really lucky, they could turn out to be right.

My investigations have focused in part on supersymmetry. If the world is indeed supersymmetric, for every known particle there is a partner antiparticle that has the same charges but different spin. The superpartners, as they are known, are expected to be heavier than the known Standard Model particles. They are expected to be light enough, however, that the LHC will have sufficient energy to produce them.

The problem is that current experimental limits are beginning to make ordinary vanilla supersymmetric models look increasingly unnatural. On top of that, the value of the Higgs mass stretches this same vanilla supersymmetry to its limits. The theory is motivated in part by justifying a light Higgs boson and somehow just doesn't want the Higgs boson to be this heavy. The simplest models predict that it's much lighter. The question becomes, does this mean that theorists who explore supersymmetry have been entirely on the wrong track, or have the models that have been proposed been too simplistic? Certainly, before throwing out such a beautiful idea, we would want to explore all the options.

Almost by accident, my collaborators Csaba Csaki and John Terning and I stumbled upon a possible explanation for both these phenomena. In our version of supersymmetry, some superpartners have big masses, whereas others do not. This turns out to be an important difference for LHC searches, since the particles that should be light are not the ones that have been well studied (at least so far).

On top of this, additional Higgs boson interactions automatically allow for a bigger Higgs mass. In fact, when we first started working on our paper, we were focused on the theoretically interesting aspect of two different scales for new particle masses. When we were finishing our paper in December, it was almost an afterthought to mention that our model readily accommodates the Higgs boson mass that was first suggested at that time. At the last minute, we remembered to change our abstract. With the Higgs mass confirmation, I'm glad we remembered.

In general, the search strategies for a modified version of a supersymmetric model—or other extensions even to known ideas—are likely to be very different. One of the chief reasons model building in particle physics is so important is to ensure that all possible signatures of interest are searched for. The number of collisions at the LHC is so enormous that unless experimenters have some definite target in mind, the process of interest can be buried in the data.

But the LHC will explore other avenues too. Dark matter

searches are on the agenda, as are searches for new heavier particles of various sorts. The Higgs boson discovery is inspiring, in that the particle was predicted and found. But most of us are humble enough to realize that nature can have surprises in store.

RELIGION, UTILITY, AND ALL THAT . . .

Since I'm addressing some of the questions I've been asked since the discovery, I'll take a moment to address three that I've heard a lot: What is this useful for? What does it tell us about religion? What do you think of the name "God particle"?

You might guess that I'm not a big fan of the name "God particle." One interviewer protested that the particle is very important and is critical to matter as we know it. Perhaps out of boredom (or maybe because I was in Greece), I pointed out that many particles are critical to matter as we know it. In a monotheistic universe it would be an overstatement to single out the Higgs boson as a deity. Maybe in a pantheistic universe, we could have many god particles. But really, they are particles and have nothing to do with religion. We build up matter at different scales, as I describe in *Knocking on Heaven's Door.* The Higgs boson is important but it has a very well-defined role.

On a related topic, the Higgs boson discovery says nothing about religion. Surprisingly, several interviewers thought it did and would even bring people to church. I would think this discovery would lend credence to the scientific method, and perhaps spark curiosity about understanding the world through science. Even those of us who trust the scientific method are very excited when a prediction proves true. After all, what the scientific method allows us to do is both rule out and verify theories by testing their experimental consequences. The Higgs boson prediction turned out to be right. This was based on theoretical considerations that took into account existing measurements. It is a tribute to science and the ingenuity of both theorists and experimenters that such a prediction could be made and verified. The discovery is truly inspirational—in a scientific way.

And is it useful? The Higgs boson has no practical implications that we know of. But believe it or not, no one knew what the electron was good for when it was first discovered. The same applies to quantum mechanics, which was critical to semiconductors and the current electronics industry. So not being able to think of practical applications isn't overly surprising.

We do know that the discovery is good for piquing humanity's curiosity and rewarding our ability to ask—and answer—deep and fundamental questions. Societies accompany advanced science with advanced education and generally with a thriving economy that derives directly and indirectly from scientific developments. After all, powerful computing methods, important magnet developments, and precision electronics all are needed to make the LHC and its experiments work. Superconducting magnet technology that was developed for accelerators is now used in medical and industrial applications. And the World Wide Web was developed at CERN to allow efficient information transfer among collaborators in different countries. Technical engineering advances—along with the mathematical and theoretical advances that inspire students and the public—help advance societies.

But really, scientists have discovered a new particle—one that tells us about the power of empty space. It was predicted almost fifty years ago based on theoretical considerations and the need to make the Standard Model consistent. It was verified through heroic engineering and experimental techniques. The particle's discovery is tremendously exciting. It's also inspirational. Let's just enjoy that for now.

The following excerpts provide more in-depth material about the Higgs mechanism and boson from *Warped Passages* and *Knocking on Heaven's Door.* Those books include still more about symmetry, particle physics, the Higgs mechanism and boson, supersymmetry, extra dimensions, LHC experiments, the nature of science, and the many elements of the scientific process.

An Excerpt from *Warped Passages:*

CHAPTER 10

THE ORIGIN OF ELEMENTARY PARTICLE MASSES: SPONTANEOUS SYMMETRY BREAKING AND THE HIGGS MECHANISM

Symmetries are important, but the universe usually doesn't manifest perfect symmetry. Slightly imperfect symmetries are what makes the world interesting (but organized). For me, one of the most intriguing aspects of physics research is the quest for connections that make symmetry meaningful in an unsymmetric world.

When a symmetry is not exact, physicists say the symmetry is broken. Although broken symmetry is often interesting, it isn't always aesthetically appealing: the beauty and economy of the underlying system or theory can be lost (or lessened). Even the very symmetric Taj Mahal doesn't have perfect symmetry, since the builder's parsimonious heir decided not to build a planned second monument, adding instead an off-center tomb to the original. This second tomb destroys the Taj Mahal's otherwise perfect fourfold rotational symmetry, detracting slightly from its underlying beauty.

But fortunately for aesthetically minded physicists, broken symmetries can be even more beautiful and interesting than things that are perfectly symmetrical. Perfect symmetry is often boring. The *Mona Lisa* with a symmetric smile just wouldn't be the same.

In physics, as in art, simplicity alone is not necessarily the highest goal. Life and the universe are rarely perfect, and almost all symmetries you care to name are broken. Although we physicists value and admire symmetry, we still have to find a connection between a symmetric theory and an asymmetric world. The best theories respect the elegance of symmetric theories while incorporating the symmetry breaking necessary to make predictions that agree with phenomena in our world. The goal is to make theories that are richer and sometimes even more beautiful without compromising their elegance.

The concept of the Higgs mechanism, which relies on the phenomenon of spontaneous symmetry breaking (which we will consider in the following section), is an example of such a sophisticated, elegant theoretical idea. This mechanism, named after the Scottish physicist Peter Higgs, lets the Standard Model particles—quarks, leptons, and weak gauge bosons—acquire mass.

Without the Higgs mechanism, all elementary particles would have to be massless; the Standard Model with massive particles but without the Higgs mechanism would make nonsensical predictions at high energies. The magical property of the Higgs mechanism is that it lets you have your cake and eat it too: particles get mass, but they act as if they are massless when they have energies at which massive particles would otherwise cause problems. We will see that the Higgs mechanism allows particles to have mass but travel freely over a restricted range, in much the same way that Ike's car, which was stopped by policemen after half a mile, traveled undisturbed over limited distances, and that this suffices to solve high-energy problems.

Although the Higgs mechanism is one of the nicest ideas in quantum field theory and underlies all fundamental particle masses, it is also somewhat abstract. For this reason it is not well known by most people aside from specialists. While you can understand many features of ideas I discuss later in the book without knowing the details of the Higgs mechanism (and you can skip now to the summary bullets if you like), this chapter does provide an opportunity to delve a bit deeper into particle physics and into the ideas, such as spontaneously symmetry breaking, that buttress theoretical developments in particle physics today. As an added bonus, some familiarity with the Higgs mechanism will let you in on an amazing insight into electromagnetism that was discovered only in the 1960s, once the weak force and the Higgs mechanism were properly understood. And later on, when we come to explore extra-dimensional models, some understanding of the Higgs mechanism will make the potential merits of those recent ideas meaningful.

SPONTANEOUSLY BROKEN SYMMETRY

Before describing the Higgs mechanism, we need first to investigate spontaneous symmetry breaking, a special type of symmetry breaking that is central to the Higgs mechanism. Spontaneous symmetry breaking plays a big role in many of the properties of the universe that we already understand and is likely to play a role in whatever we have yet to discover.

Spontaneous symmetry breaking is not only ubiquitous in physics, but is a prevalent feature of everyday life. A spontaneously broken symmetry is a symmetry that is preserved by physical laws but not by the way things are actually arranged in the world. Spontaneous symmetry breaking takes place when a system cannot preserve a symmetry that would otherwise be present. Perhaps the best way to explain how this works is to give a few examples.

Let's first consider a dinner arrangement in which a number of people are seated around a circular table with water glasses placed between them. Which glass should someone use, the one on the right or the one on the left? There is no good answer. I am told that Miss Manners says the one on the right, but aside from arbitrary rules of etiquette, left and right serve equally well.

However, as soon as someone chooses a glass, the symmetry is broken. The impetus to choose would not necessarily be part of the system; in this case it would be another factor—thirst. Nonetheless, if one person spontaneously drank from the glass on their left, so would that person's neighbors, and in the end everyone would have drunk from the glass on the left.

The symmetry exists until the moment someone picks up a glass. At that moment the left-right symmetry is spontaneously broken. No law of physics dictates that anyone has to choose left or right. But one has to be chosen, and after that, left and right are no longer the same in that there is no longer a symmetry that interchanges the two.

Here's another example. Imagine a pencil standing on end at the center of a circle. For the split second in which it rests on its tip and is exactly vertical, all directions are equivalent and a rotational symmetry exists. But a pencil standing on end won't just stay there: it will spontaneously fall in some direction. As soon as the pencil topples over, the original rotational symmetry is broken.

Notice that it would not be the physical laws themselves that determined the direction. The physics of the pencil falling over would be exactly the same no matter the direction in which it fell. What would break the symmetry would be the pencil itself, the state of the system. The pencil simply cannot fall in all directions at once. It has to fall in one particular direction.

A wall that is infinitely long and high would also look the same everywhere and in all directions along it. But because an actual wall has boundaries, if you are to see the symmetries you will have to get close enough to it that the boundaries are out of your field of vision. The wall's ends tell you that not everywhere along the wall is the same, but if you were to press your nose up against it so that you could see only a short distance away, the symmetry would appear to be preserved. You might want to briefly reflect on this example, which shows that a symmetry can appear to be preserved when viewed on one distance scale, even though it appears to be broken on another—a concept whose importance will become apparent very soon.

Almost any symmetry you care to name is not preserved in

the world. For example, there are many symmetries that would be present in empty space, such as rotational or translational invariance, which tell us that all directions and positions are the same. But space is not empty: it is punctuated by structures such as stars and the solar system, which occupy particular positions and are oriented in particular ways that no longer preserve the underlying symmetry. They could have been anywhere, but they can't be everywhere. The underlying symmetries must be broken, although they remain implicit in the physical laws describing the world.

The symmetry associated with the weak force is also spontaneously broken. In the rest of this chapter I'll explain how we know this and discuss some of the consequences. We'll see that spontaneously breaking the weak force symmetry is the only way to explain massive particles while avoiding incorrect predictions for high-energy particles that cannot be avoided in any other candidate theory. The Higgs mechanism acknowledges both the requirement of an internal symmetry associated with the weak force and the necessity for it being broken.

THE PROBLEM

The weak force has one especially bizarre property. Unlike the electromagnetic force, which travels over large distances—which you benefit from each time you turn on the radio—the weak force affects only matter that is within extremely close range. Two particles must be within one ten thousand trillionth of a centimeter to influence each other via the weak force.

For the physicists who studied quantum field theory and quantum electrodynamics (QED, the quantum field theory of electromagnetism) in its earliest days, this restricted range was a mystery. QED made it look as if forces, such as the well-understood electromagnetic force, should be transmitted arbitrarily far away from a charged source. Why wasn't the weak force also communicated to particles at any distance and not just to those nearby?

Quantum field theory, which combines the principles of quantum mechanics and special relativity, dictates that if low-energy particles communicate forces only a short distance, they must have mass; and the heavier the particle, the shorter the particle's range. As explained in Chapter 6, this is a consequence of the uncertainty principle and special relativity. The uncertainty principle tells us that you need high-momentum particles to probe or influence physical processes at short distances, and special relativity relates that momentum to a mass. Although this is a qualitative statement, quantum field theory makes this relationship precise. It tells how far a massive particle will travel: the smaller the mass, the bigger the distance.

Therefore, according to quantum field theory, the short range of the weak force could mean only one thing: the weak gauge bosons communicating the force had to have nonzero mass. However, the theory of forces I described in the previous chapter works only for gauge bosons such as the photon, which communicates a force over large distances and has zero mass. According to the original theory of forces, the existence of nonzero masses was strange and problematic—the theory's high-energy predictions when gauge bosons have mass make no sense. For example, the theory would predict that very energetic, massive gauge bosons would interact much too strongly—so strongly in fact that particles would appear to be interacting more than 100 percent of the time. This naive theory is clearly wrong.

Furthermore, the masses for weak gauge bosons, quarks, and leptons (all of which we know to have nonzero mass) do not preserve the internal symmetry which, as we saw in the previous chapter, is a key ingredient in the theory of forces. Physicists who hoped to construct a theory with massive particles clearly needed a new idea.

Physicists have shown that the only way to make a theory that avoids nonsensical predictions about energetic, massive gauge bosons is to have the weak force symmetry break spontaneously through the process known as the Higgs mechanism. Here's why.

You might recall from the previous chapter that one of the reasons we wanted to include an internal symmetry that eliminates one of the three possible polarizations of a gauge boson was that a theory without the symmetry makes the same sort of nonsensical predictions I've just mentioned. The simplest theory of forces without an internal symmetry predicts that any energetic gauge boson, with or without a mass, interacts with other gauge bosons far too often.

The successful theory of forces eliminates this bad high-energy behavior by forbidding the polarization that is responsible for the incorrect predictions and doesn't actually exist in nature. Spurious polarizations are the source of the problematic predictions for high-energy scattering, so the symmetry allows only physical polarizations—the ones that really exist and are consistent with the symmetry—to remain. The symmetry, which rids the theory of nonexistent polarizations, also eliminates the incorrect predictions they would otherwise induce.

Although I didn't say so explicitly at the time, this idea works as stated only for massless gauge bosons. The weak gauge bosons, unlike the photon, have nonzero masses. Weak gauge bosons travel at less than the speed of light. And that puts a wrench in the works.

Whereas massless gauge bosons have only two polarizations that exist in nature, massive gauge bosons have three. One way to understand this distinction is that massless gauge bosons always travel at the speed of light, which tells us that they are never at rest. They therefore always single out their direction of motion, so you can always distinguish the perpendicular directions from the remaining polarization along the direction of travel. And it turns out that for massless gauge bosons, physical polarizations oscillate only in the two perpendicular directions.

Massive gauge bosons, on the other hand, are different. Like all familiar objects, they can sit still. But when a massive gauge boson isn't moving, it doesn't single out a direction of motion. To a massive gauge boson sitting at rest, all three directions should be equivalent. But if all three directions are equivalent, then all three possible polarizations would have to exist in nature. And they do.

Even if you find the above logic mysterious, rest assured that experimenters have already seen the effects of a third polarization of a massive gauge boson and have confirmed its existence. The third polarization is called the longitudinal polarization. When a massive gauge boson is moving, the longitudinal polarization is the wave that oscillates along the direction of motion—the direction of sound wave oscillations, for example.

This polarization doesn't exist in the case of massless gauge bosons such as the photon. However, for massive gauge bosons, like the weak gauge bosons, the third polarization is truly a part of nature. This third polarization must be a part of the weak gauge boson theory.

Because this third polarization is the source of the weak gauge boson's overly large interaction rate at high energy, its existence poses a dilemma. We already know that we need a symmetry to eliminate the bad high-energy behavior. But this symmetry gets rid of the incorrect predictions by eliminating the third polarization as well, and that polarization is essential to a massive gauge boson and therefore to the theory that describes it. Although an internal symmetry would eliminate bad predictions for high-energy behavior, it would do so at too high a price: the symmetry would get rid of the mass as well! A symmetry in the theory of massive gauge bosons seems poised to throw away the baby with the bathwater.

The impasse at first glance looks insurmountable, since the requirements for a theory of massive gauge bosons appear to be entirely contradictory. On the one hand, an internal symmetry—the one described in the previous chapter—should not be preserved, since otherwise massive gauge bosons with three physical polarizations would be forbidden. On the other hand, without an internal symmetry to eliminate two of the polarizations, the theory of forces makes incorrect predictions when the gauge bosons have high energy. We still need a symmetry to eliminate the third polarization of each massive gauge boson if we are to have any hope of eliminating the bad high-energy behavior.

The key to resolving this apparent paradox and figuring out the correct quantum field theory description of a massive gauge boson was recognizing the difference between the ones with high energy and the ones with low energy. In the theory without an internal symmetry, only predictions about the high-energy gauge bosons looked as if they would be problematic. Predictions about low-energy massive gauge bosons were sensible (and true).

These two facts together imply something fairly profound: to avoid problematic high-energy predictions, an internal symmetry is essential—the lessons of the previous chapter still apply. But when the massive gauge bosons have low energy (low compared with the energy that Einstein's relation $E = mc^2$ associates with its mass), the symmetry should no longer be preserved. The symmetry must be eliminated so that gauge bosons can have mass and the third polarization can participate in the low-energy interactions where the mass makes a difference.

In 1964, Peter Higgs and others discovered how theories of forces could incorporate massive gauge bosons by doing exactly what we just said: keeping an internal symmetry at high energies, but eliminating it at low energies. The Higgs mechanism, based on spontaneous symmetry breaking, breaks the internal symmetry of the weak interactions, but only at low energy. That ensures that the extra polarization will be present at low energy, where the theory needs it. But the extra polarization will not participate in high-energy processes, and the nonsensical high-energy interactions will not appear.

Let's now consider a particular model that spontaneously breaks the weak force symmetry and implements the Higgs mechanism. With this exemplar of the Higgs mechanism, we'll see how the elementary particles of the Standard Model acquire mass.

THE HIGGS MECHANISM

The Higgs mechanism involves a field that physicists call the Higgs field. As we have seen, the fields of quantum field theory are objects that can produce particles anywhere in space. Each type of field generates its own particular type of particle. An electron field is the source of electrons, for example. Similarly, a Higgs field is the source of Higgs particles.

As with heavy quarks and leptons, Higgs particles are so heavy that they aren't found in ordinary matter. But unlike heavy quarks and leptons, no one has ever observed the Higgs particles that the Higgs field would produce, even in experiments performed at high-energy accelerators. This doesn't mean that Higgs particles don't exist, just that Higgs particles are too heavy to have been produced with the energies that experiments have explored so far. Physicists expect that if Higgs particles exist, we'll create them in only a few years' time, when the higher-energy LHC collider comes into operation.

Nevertheless, we are fairly confident the Higgs mechanism applies to our world, since it is the only known way to give Standard Model particles their masses. It is the only known solution to the problems that were posed in the previous section. Unfortunately, because no one has yet discovered the Higgs particle, we still don't know precisely what the Higgs field (or fields) actually is.

The nature of the Higgs particle is one of the most hotly debated topics in particle physics. In this section, I will present the simplest of many candidate models—possible theories that contain different particles and forces—that demonstrates how the Higgs mechanism works. Whatever the true Higgs field theory turns out to be, it will implement the Higgs mechanism—spontaneously breaking the weak force symmetry and giving masses to elementary particles—in the same manner as the model I'm about to present.

In this model, a pair of fields experience the weak force. It will be useful later to think of these two Higgs fields, which are subject to the weak force, as carrying weak force charge. The Higgs mechanism terminology is sometimes sloppy, with "the Higgs" sometimes denoting the two fields together, and at other times one of the individual fields (and often the Higgs particles we hope to find). Here I will distinguish the possibilities and refer to the individual fields as $Higgs_1$ and $Higgs_2$.

Both $Higgs_1$ and $Higgs_2$ have the potential to produce particles. But they can also take nonzero values even when no particles are present. We haven't encountered such nonzero values for quantum fields up to this point. So far, aside from the electric and magnetic fields, we have considered only quantum fields that create or destroy particles but take zero value in the absence of particles. But quantum fields can also have nonzero values, just like the classical electric and magnetic fields. And according to the Higgs mechanism, one of the Higgs fields takes a nonzero value. We will now see that this nonzero value is ultimately the origin of particle masses.[17]

When a field takes a nonzero value, the best way to think about it is to imagine space manifesting the charge that the field carries, but not containing any actual particles. You should think of the charge that the field carries as being present everywhere. This is, alas, a rather abstract notion because the field itself is an abstract object. But when the field takes a nonzero value, its consequences are concrete: the charge that a nonzero field would carry exists in the real world.

A nonzero Higgs field, in particular, distributes weak charge throughout the universe. It is as if the nonzero weak-charge-carrying Higgs field paints weak charge throughout space. A nonzero value for the Higgs fields means that the weak charge that $Higgs_1$ (or $Higgs_2$) carries is everywhere, even when there are no particles present. The vacuum—the state of the universe with no particles present—itself carries weak charge when one of the two Higgs fields takes a nonzero value.

Weak gauge bosons interact with this weak charge of the vacuum, just as they do with all weak charge. And the charge that pervades the vacuum blocks the weak gauge bosons as they try to communicate forces over long distances. The further they try to travel, the more "paint" they encounter. (Because the charge actually spreads throughout three dimensions, you might prefer to imagine a fog of paint.)

The Higgs field plays a role very similar to that of the traffic cop in the story, restricting the weak force's influence to very short distances. When attempting to communicate the weak force to distant particles, the force-carrying weak gauge bosons bump into the Higgs field, which gets in their way and cuts them off. Like Ike, who could travel freely only within a half-mile radius of his starting point, weak gauge bosons move unimpeded only for a very short distance, about one ten thousand trillionth of a centimeter. Both weak gauge bosons and Ike are free to travel short distances, but are intercepted at longer distances.

The weak charge in the vacuum is spread out so thinly that at short distance there is very little sign of the nonzero Higgs field and the associated charge. Quarks, leptons, and the weak gauge bosons travel freely over short distances, almost as if the charge in the vacuum didn't exist. The weak gauge bosons can therefore communicate forces over short distances, almost as if the two Higgs fields were both zero.

However, at longer distances, particles travel further and therefore encounter a more significant amount of weak charge. How much they encounter depends on the charge density, which depends in turn on the value of the nonzero Higgs field. Long-distance travel (and communication of the weak force) is not an option for low-energy weak gauge bosons, for on long-distance excursions the weak charge in the vacuum gets in the way.

This is exactly what we needed to make sense of weak gauge bosons. Quantum field theory says that particles that travel freely over short distances, but only extremely rarely travel over longer distances, have nonzero masses. The weak gauge bosons' interrupted travel tells us that they act as if they have mass, since massive gauge bosons just don't get very far. The weak charge that permeates space hinders the weak gauge bosons' travel, making them behave exactly as they should in order to agree with experiments.

The weak charges in the vacuum have a density that corresponds roughly to charges that are separated by one ten thousand trillionth of a centimeter. With this weak charge density, the masses of the weak gauge bosons—the charged Ws and the neutral Z—take their measured values of approximately 100 GeV.

And that's not all that the Higgs mechanism accomplishes.

It is also responsible for the masses of quarks and leptons, the elementary particles constituting the matter of the Standard Model. Quarks and leptons acquire mass in a very similar fashion to the weak gauge bosons. Quarks and leptons interact with the Higgs field distributed throughout space and are therefore also hindered by the universe's weak charge. Like weak gauge bosons, quarks and leptons acquire mass by bouncing off the Higgs charge distributed everywhere throughout spacetime. Without the Higgs field, these particles would also have zero mass. But once again, the nonzero Higgs field and the vacuum's weak charge interfere with motion and make the particles have mass. The Higgs mechanism is also necessary for quarks and leptons to acquire their masses.

Although the Higgs mechanism is a more elaborate origin of mass than you might think necessary, it is the only sensible way for the weak gauge bosons to acquire mass according to quantum field theory. The beauty of the Higgs mechanism is that it gives the weak gauge bosons mass while accomplishing precisely the task I laid out at the beginning of this chapter. The Higgs mechanism makes it look as though the weak force symmetry is preserved at short distances (which, according to quantum mechanics and special relativity, is equivalent to high energy) but is broken at long distances (equivalent to low energy). It breaks the weak force symmetry spontaneously, and this spontaneous breaking lies at the root of the solution to the problem of massive gauge bosons. This more advanced topic is explained in the following section (but feel free to skip ahead to the following chapter if you wish).

SPONTANEOUS BREAKING OF WEAK FORCE SYMMETRY

We have seen that the internal symmetry transformation associated with the weak force will interchange anything that is charged under the weak force because the symmetry transformation acts on anything that interacts with weak gauge bosons. Therefore, the internal symmetry associated with the weak force must act on the Higgs_1 and Higgs_2 fields, or the Higgs_1 and Higgs_2 particles they would create, and treat them as equivalent, just as it treats up and down quarks, which also experience the weak force, as interchangeable particles.

If both of the Higgs fields were zero, they would be equivalent and interchangeable, and the full symmetry associated with the weak force would be preserved. However, when one of the two Higgs fields takes a nonzero value, the Higgs fields spontaneously break the symmetry of the weak force. If one field is zero and the other is not, the electroweak symmetry, by which Higgs_1 and Higgs_2 are interchangeable, is broken.

Just as the first person to choose his left or right glass breaks the left-right symmetry at a round table, one Higgs field taking a nonzero value breaks the weak force symmetry that interchanges the two Higgs fields. The symmetry is broken spontaneously because all that breaks it is the vacuum—the actual state of the system, the nonzero field in this case. Nonetheless, the physical laws, which are unchanged, still preserve the symmetry.

A picture might help convey how a nonzero field breaks the weak force symmetry. Figure 58 shows a graph with two axes, labeled x and y. The equivalence of the two Higgs fields is like the equivalence of the x and y axes with no points plotted. If I were to rotate the graph so that the axes switched places, the picture would still look the same. This is a consequence of ordinary rotational symmetry.[18]

Notice that if I plot a point at the position x = 0, y = 0, this rotational symmetry is completely preserved. But if I plot a point that has one nonzero coordinate value, for example where x = 5 and y = 0, the rotational symmetry is no longer preserved. The two axes are no longer equivalent because the x value, but not the y value, of this point is not zero.[19]

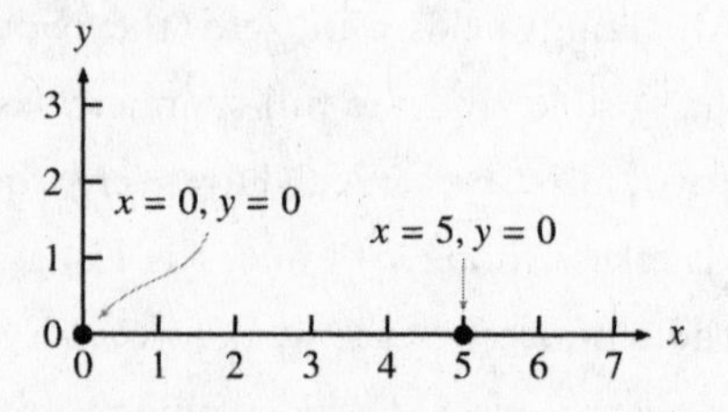

Figure 58. When the point x = 0, y = 0 is singled out, rotational symmetry is preserved. But when x = 5, y = 0 is singled out, rotational symmetry is broken.

The Higgs mechanism spontaneously breaks weak force symmetry in a similar fashion. When the two Higgs fields are zero, the symmetry is preserved. But when one is zero and the other is not, the weak force symmetry is spontaneously broken.

The weak gauge boson masses tell us the precise value of the energy at which the weak force symmetry is spontaneously broken. That energy is 250 GeV, the weak scale energy, very close to the masses of the weak gauge bosons, the W^-, the W^+ and the

Z. When particles have energy greater than 250 GeV, interactions occur as if the symmetry is preserved, but when their energy is less than 250 GeV, the symmetry is broken and weak gauge bosons act as if they have mass. With the correct value of the nonvanishing Higgs field, the weak force symmetry is spontaneously broken at the right energy, and the weak gauge bosons get precisely the right mass.

The symmetry transformations that act on the weak gauge bosons also act on quarks and leptons. And it turns out that these transformations won't leave things the same unless quarks and leptons are massless. That means that weak force symmetries would be preserved only if quarks and leptons didn't have mass. And because the weak force symmetry is essential at high energies, not only is spontaneous symmetry breaking required for the weak gauge boson masses, it's necessary for the quarks and lepton to acquire masses as well. The Higgs mechanism is the only way for all the massive fundamental particles of the Standard Model to acquire their masses.

The Higgs mechanism functions in exactly the way that is needed to ensure that any theory that incorporates it can have massive weak gauge bosons (as well as massive quarks and leptons) and nonetheless will make the correct predictions for high-energy behavior. Specifically, for high-energy weak gauge bosons—those with energy larger than 250 GeV—symmetry appears to be preserved, so there are no incorrect predictions. At high energies the internal symmetry associated with the weak force still filters out the problematic polarization of the weak gauge boson that would cause interactions at too high a rate. But at low energies, where the mass is essential to reproducing the measured short-range interactions of the weak force, the weak force symmetry is broken.

This is why the Higgs mechanism is so important. No other theory that assigns these masses has these properties. Other approaches fail either at low energies, where the mass will be wrong, or at high energies, where interactions will be predicted incorrectly.

BONUS

There is one more successful feature of the Standard Model that I have not yet explained. Although the Higgs field will be relevant to the next few chapters, this particular aspect of the Higgs mechanism will not. Yet it is so surprising and fascinating that it's worth mentioning.

The Higgs mechanism tells us about more than just the weak force. Surprisingly, it also gives new insight into why electromagnetism is special. Until the 1960s, no one would have believed that there was more to learn about the electromagnetic force, which had been so well studied and understood for over a century. In the 1960s, however, the electroweak theory proposed by Sheldon Glashow, Steven Weinberg, and Abdus Salam showed that when the universe began its evolution at high temperature and energy, there were three weak gauge bosons, plus a fourth, independent, neutral boson with a different interaction strength. The photon, ubiquitous and important as it is today, was not a member of this list. The authors of the electroweak theory deduced the nature of the four weak gauge bosons from both mathematical and physical clues, which I won't go into here.

The remarkable thing is that the photon was originally nothing special. In fact, the photon we talk about today is actually a mixture of two of the original four gauge bosons. The reason that the photon got singled out is that it is the only gauge boson involved in the electroweak force that is impervious to the weak charge of the vacuum. The chief distinguishing feature of the photon is that it travels unfettered through the weakly charged vacuum and therefore has no mass.

Photon travel, unlike that of the W and Z, is not obstructed by the nonzero value of a Higgs field. That's because although the vacuum carries weak charge, it does not carry electric charge. The photon, which communicates the electromagnetic force, interacts only with electrically charged objects. For this reason, the photon can communicate a long-range force without any interference from the vacuum. It is therefore the only gauge boson that remains massless even in the presence of the nonzero Higgs field.

The situation closely resembles the speed traps with which Ike had to contend (although this part of the analogy is admittedly a little more tenuous). The speed traps let dull cars pass through scot-free. Photons, like the dull neutral cars, always travel unimpeded.

Who would have thought it? The photon, the thing that physicists for years thought they understood completely, has an origin that can be understood only in terms of a more complex theory that combines the weak and electromagnetic forces into a single theory. This theory is therefore generally referred to as the electroweak theory, and the relevant symmetry is called electroweak symmetry. The electroweak theory and the Higgs mechanism are major successes of particle physics. Not only the weak gauge boson masses, but also the relevance of the photon are neatly explained within its framework. On top of that, it allows us to understand the origin of the quark and lepton masses. The rather abstract ideas we have just encountered nicely explain quite a wide range of features of the world.

CAVEAT

The Higgs mechanism works beautifully, and gives quarks, leptons, and weak gauge bosons their masses without making nonsensical high-energy predictions—and, furthermore, explains how the photon came to be. However, there is one essential property of the Higgs particle that physicists don't yet fully understand.

Electroweak symmetry must be broken at about 250 GeV to give particles their measured masses. Experiments show that particles with energy greater than 250 GeV look as if they are massless, whereas particles with energy below 250 GeV act as if they have mass. However, the electroweak symmetry will break at 250 GeV only if the Higgs particle (sometimes also called the Higgs boson)[20] itself has roughly this mass (again, using $E = mc^2$); the theory of the weak force wouldn't work if the Higgs mass were much greater. If the Higgs mass were greater, symmetry breaking would happen at a higher energy and the weak gauge bosons would be heavier—contradicting experimental results.

However, in Chapter 12, I will explain why a light Higgs

particle poses a major theoretical problem. Calculations that take quantum mechanics into account point to a heavier Higgs particle, and physicists don't yet know why the Higgs particle mass should be so low. This quandary is critical to motivating new particle physics ideas and some of the extra-dimensional models that we'll consider later on.

Even without knowing the precise nature of the Higgs particle and the reason why it is so light, the mass requirement tells us that the Large Hadron Collider, which will start operating at CERN in Switzerland within the decade, should discover one or more crucial new particles. Whatever breaks electroweak symmetry must have a mass that is around the weak scale mass. And we expect that the LHC will find out what it is. When it does, this critically important discovery will greatly advance our knowledge of the underlying structure of matter. And it will also tell us which (if any) of the proposals for explaining the Higgs particle is correct.

But before we get to those proposals, we'll look at one possible extension of the Standard Model that was suggested purely in the interest of simplicity of nature. The next chapter explores virtual particles, the distance dependence of forces, and the fascinating topic of grand unification.

WHAT TO REMEMBER

* Despite the importance of symmetries for making the right predictions about high-energy particles, the masses of quarks, leptons, and weak gauge bosons tell us that the weak force symmetry must be broken.

* Because we still have to guard against false predictions, the weak force symmetry must nonetheless be maintained at high energy. Therefore, the weak force symmetry must be broken only at low energy.
* Spontaneous symmetry breaking occurs when all physical laws preserve a symmetry but the actual physical system does not. Spontaneously broken symmetries are symmetries that are preserved at high energies but broken at low energies. The weak force symmetry is spontaneously broken.
* The process by which weak force symmetry is spontaneously broken is the Higgs mechanism. For the Higgs mechanism to spontaneously break the weak force symmetry, there has to be a particle with a mass of about the weak scale mass, which is 250 GeV (remember, special relativity relates energy and mass through $E = mc^2$).

An Excerpt from *Knocking on Heaven's Door:*

CHAPTER 16

THE HIGGS BOSON

On the morning of March 30, 2010, I awoke to a flurry of e-mails about the successful 7 TeV collisions that had taken place at CERN the night before. This triumph launched the beginning of the true physics program at the LHC. The acceleration and collisions that had taken place toward the close of the previous year had been critical technical milestones. Those events were important for LHC experimenters who could finally calibrate and better understand their detectors using data from genuine LHC collisions, and not just cosmic rays that had happened to pass through their apparatus. But for the next year and a half, detectors at CERN would be recording real data that physicists could use to constrain or verify models. Finally, after its many ups and downs, the physics program at the LHC had at long last begun.

The launch proceeded almost exactly according to plan—a good thing according to my experimental colleagues, who the day before had expressed concerns that the presence of reporters might compromise the day's technical goals. The reporters (and everyone else present) did witness a couple of false starts—in part because of the zealous protection mechanisms that had been installed, which were designed to trigger if anything went even slightly awry. But within a few hours, beams circulated and collided and newspapers and websites had lots of pretty pictures to display.

The 7 TeV collisions occurred with only half the intended LHC energy. The real target energy of 14 TeV wouldn't be reached for several years. And the intended luminosity for the 7 TeV run—the number of protons that would collide each second—was much lower than designers had originally planned. Still, with these collisions, everything at the LHC was at long last on track. We could finally believe that our understanding of the inner nature of matter would soon improve. And if all went okay, in a couple of years the machine would shut down, gear up, and come back online at full capacity and provide the real answers we were waiting for.

One of the most important goals will be learning how fundamental particles acquire their mass. Why isn't everything whizzing around at the speed of light, which is what matter would do if it had zero mass? The answer to this question hinges on the set of particles that are known collectively as the *Higgs sector*, including the *Higgs boson*. This chapter explains why a successful search for this particle will tell us whether our ideas about how elementary particle masses arise are correct. Searches that will take place once the LHC comes back online with higher intensity and greater energy should ultimately tell us about the particles and interactions that underlie this critical and rather remarkable phenomenon.

THE HIGGS MECHANISM

No physicist questions that the Standard Model works at the energies we have studied so far. Experiments have tested its many predictions, which agree with expectations to better than one percent precision.

However, the Standard Model relies on an ingredient that no one has yet observed. The Higgs mechanism, named after the British physicist Peter Higgs, is the only way we know to consistently give elementary particles their mass. According to the basic premises of the naive version of the Standard Model, neither the gauge bosons that communicate forces nor the elementary particles, such as quarks and leptons that are essential to the Standard Model should have nonzero masses. Yet measurements of physical phenomena clearly demonstrate that they do. Elementary particle masses are critical to understanding atomic and particle physics phenomena, such as the radius of an electron's orbit in an atom or the extremely tiny range of the weak force, not to mention the formation of structure in the universe. Masses also determine how much energy is needed to create elementary particles—in accordance with the equation $E = mc^2$. Yet in the Standard Model without a Higgs mechanism, elementary particles' masses would be a mystery. They would not be allowed.

The notion that particles don't have an inalienable right to their masses might sound needlessly autocratic. You could quite reasonably expect that particles always have the option of possessing a nonvanishing mass. Yet the subtle structure of the Standard Model and any theory of forces is just that tyrannical. It constrains the types of masses that are allowed. The explanations will seem a little different for gauge bosons than for fermions, but the underlying logic for both relates to the symmetries at the heart of any theory of forces.

The Standard Model of particle physics includes the electromagnetic, weak, and strong nuclear forces, and each force is associated with a symmetry. Without such symmetries, too many oscillation modes of the gauge bosons—the particles that communicate those forces—would be predicted to be present by the theory that quantum mechanics and special relativity tells us describes them. In the theory without symmetries, theoretical calculations would generate nonsensical predictions, such as probabilities for high-energy interactions greater than one for the spurious oscillation modes. In any accurate description of nature, such unphysical particles—particles that don't actually exist because they oscillate in the wrong direction—clearly need to be eliminated.

In this context, symmetries act like spam filters, or quality control constraints. Quality requirements might specify keeping only those cars that are symmetrically balanced, for instance, so that the cars that make it out of the factory all behave as expected. Symmetries in any theory of forces also screen out the badly behaved elements. That's because interactions among the undesirable, unphysical particles don't respect the symmetries, whereas those particles that interact in a way that preserves the necessary symmetries oscillate as they should. Symmetries thereby guarantee that theoretical predictions involve only the physical particles and therefore make sense and agree with experiments.

Symmetries therefore permit an elegant formulation of a theory of forces. Rather than eliminate unphysical modes in each calculation one by one, symmetries eliminate all the unphysical particles with one fell swoop. Any theory with symmetric interactions involves only the physical oscillation modes whose behavior we want to describe.

This works perfectly in any theory of forces involving zero mass force carriers, such as electromagnetism or the strong nuclear force. In symmetric theories, predictions for their high-energy interactions all make sense and only physical modes—modes that exist in nature—get included. For massless gauge bosons, the problem with high-energy interactions is relatively straightforward to solve, since appropriate symmetry constraints remove any unphysical, badly behaved modes from the theory.

Symmetries thereby solve two problems: unphysical modes are eliminated, and the bad high-energy predictions that would accompany them are as well. However, a gauge boson with nonzero mass has an additional physical—existent in nature—mode of oscillation. The gauge bosons that communicate the weak nuclear force fall into this category. Symmetries would eliminate too many of their oscillation modes. Without some new ingredient, weak boson masses cannot respect the Standard Model symmetries. For gauge bosons with nonzero mass, we have no choice but to keep a badly behaved mode—and that means the solution to the bad high-energy behavior is not so simple. Nonetheless something is still required for the theory to generate sensible high-energy interactions.

Moreover, none of the elementary particles in the Standard Model without a Higgs can have a nonzero mass that respects the symmetries of the most naive theory of forces. With the symmetries associated with forces present, quarks and leptons in the

Higgsless Standard Model would not have nonzero masses either. The reason appears to be unrelated to the logic about gauge bosons, but it can also be traced to symmetries.

In Chapter 14, we presented a table that included both left- and right-handed fermions—particles that get paired in the presence of nonzero masses. When quark or lepton masses are nonzero, they introduce interactions that convert left-handed fermions to right-handed fermions. But for left-handed and right-handed fermions to be interconvertible, they would both have to experience the same forces. Yet experiments demonstrate that the weak force acts differently on left-handed fermions than on the right-handed fermions that massive quarks or leptons could turn into. This violation of parity symmetry, which if preserved would treat left and right as equivalent for the laws of physics, is startling to everyone when they first learn about it. After all, the other known laws of nature don't distinguish left and right. But this remarkable property—that the weak force does not treat left and right the same—has been demonstrated experimentally and is an essential feature of the Standard Model.

The different interactions of left- and right-handed quarks and leptons tells us that without some new ingredient, nonzero masses for quarks and leptons would be inconsistent with known physical laws. Such nonzero masses would connect particles that carry weak charge with particles that do not.

In other words, since only left-handed particles carry this charge, weak charge could be lost. Charges would apparently disappear into the *vacuum*—the state of the universe that doesn't contain any particles. Generally that should not happen. Charges should be conserved. If charge could appear and disappear, the symmetries associated with the corresponding force would be broken, and the bizarre probabilistic predictions about high-energy gauge boson interactions that they are supposed to eliminate would reemerge. Charges should never magically disappear in this manner if the vacuum is truly empty and contains no particles or fields.

But charges can appear and disappear if the vacuum is not really empty—but instead contains a *Higgs field* that supplies weak charge to the vacuum. A Higgs field, even one that gives charge to the vacuum, isn't composed of actual particles. It is essentially a distribution of weak charge throughout the universe that happens only when the field itself takes a nonzero value. When the Higgs field is nonvanishing, it is as if the universe has an infinite supply of weak charges. Imagine that you had an infinite supply of money. You could lend or take money at will and you would always still have an infinite amount at your disposal. In a similar spirit, the Higgs field puts infinite weak charge into the vacuum. In doing so, it breaks the symmetries associated with forces and lets charges flow into and out of the vacuum so that particle masses arise without causing any problems.

One way to think about the Higgs mechanism and the origin of masses is that it lets the vacuum behave like a viscous fluid—a Higgs field that permeates the vacuum—that carries weak charge. Particles that carry this charge, such as the weak gauge bosons and Standard Model quarks and leptons, can interact with this fluid, and these interactions slow them down. This slowing down then corresponds to the particles acquiring mass, since particles without mass will travel through the vacuum at the speed of light.

This subtle process by which elementary particles acquire

their masses is known as the Higgs mechanism. It tells us not only how elementary particles acquire their masses, but also quite a bit about those masses' properties. The mechanism explains, for instance, why some particles are heavy while others are light. It is simply that particles that interact more with the Higgs field have larger masses and those that interact less have smaller ones. The top quark, which is the heaviest, has the biggest such interaction. An electron or an up quark, which have relatively small masses, have much more feeble ones.

The Higgs mechanism also provides a deep insight into the nature of electromagnetism and the photon that communicates that force. The Higgs mechanism tells us that only those force carriers that interact with the weak charge distributed throughout the vacuum acquire mass. Because the *W* gauge bosons and the *Z* boson interact with these charges, they have nonvanishing masses. However, the Higgs field that suffuses the vacuum carries weak charge but is electrically neutral. The photon doesn't interact with the weak charge, so its mass remains zero. The photon is thereby singled out. Without the Higgs mechanism, there would be three zero mass weak gauge bosons and one other force carrier—also with zero mass—known as the hypercharge gauge boson. No one would ever mention a photon at all. But in the presence of the Higgs field, only a unique combination of the hypercharge gauge boson and one of the three weak gauge bosons will not interact with the charge in the vacuum—and that combination is precisely the photon that communicates electromagnetism. The photon's masslessness is critical to the important phenomena that follow from electromagnetism. It explains why radio waves can extend over enormous distances, while the weak force is screened over extremely tiny ones. The Higgs field carries weak charge—but no electric charge. So the photon has zero mass and travels at the speed of light—by definition—while the weak force carriers are heavy.

Don't be confused. Photons are elementary particles. But in a sense, the original gauge bosons were misidentified since they didn't correspond to the physical particles that have definite masses (which might be zero) and travel through the vacuum unperturbed. Until we know the weak charges that are distributed throughout the vacuum via the Higgs mechanism, we have no way to pick out which particles have nonzero mass and which of them don't. According to the charges assigned to the vacuum by the Higgs mechanism, the hypercharge gauge boson and the weak gauge boson would flip back and forth into each other as they travel through the vacuum and we couldn't assign either one a definite mass. Given the vacuum's weak charge, only the photon and the *Z* boson travel without changing identity as they travel through the vacuum, with the *Z* boson acquiring mass, whereas the photon does not. The Higgs mechanism thereby singles out the particular particle called the photon and the charge that we know as the electric charge which it communicates.

So the Higgs mechanism explains why it is the photon and not the other force carriers that has zero mass. It also explains one other property of masses. This next lesson is even a bit more subtle, but gives us deep insights into why the Higgs mechanism allows masses that are consistent with sensible high-energy predictions. If we think of the Higgs field as a fluid, we can imagine that its density is also relevant to particle masses. And if we think of this density as arising from charges with a fixed spacing, then these particles—which travel such small distances that they never hit a weak charge—will travel as if they had zero mass, whereas particles that travel over larger distances would inevitably bounce off weak charges and slow down.

This corresponds to the fact that the Higgs mechanism is associated with *spontaneous breaking* of the symmetry associated with the weak force—and that symmetry breaking is associated with a definite scale.

Spontaneous breaking of a symmetry occurs when the symmetry itself is present in the laws of nature—as with any theory of forces—but is broken by the actual state of a system. As we've argued, symmetries must exist for reasons connected to the high-energy behavior of particles in the theory. The only solution then is that the symmetries exist—but they are spontaneously broken so that the weak gauge bosons can have mass, but not exhibit bad high-energy behavior.

The idea behind the Higgs mechanism is that the symmetry is indeed part of the theory. The laws of physics act symmetrically. But the actual state of the world doesn't respect the symmetry. Think of a pencil that originally stood on end and then falls down and chooses one particular direction. All of the directions around the pencil were the same when it was upright, but the symmetry is broken once the pencil falls. The horizontal pencil thereby spontaneously breaks the rotational symmetry that the upright pencil preserved.

The Higgs mechanism similarly spontaneously breaks weak force symmetry. This means that the laws of physics preserve the symmetry, but it is broken by the state of the vacuum that is suffused with weak force charge. The Higgs field, which permeates the universe in a way that is not symmetric, allows elementary particles to acquire mass, since it breaks the weak force symmetry that would be present without it. The theory of forces preserves a symmetry associated with the weak force, but that symmetry is broken by the Higgs field that suffuses the vacuum.

By putting charge into the vacuum, the Higgs mechanism breaks the symmetry associated with the weak force. And it does so at a particular scale. The scale is set by the distribution of charges in the vacuum. At high energies, or equivalently—via quantum mechanics—small distances, particles won't encounter any weak charge and therefore behave as if they have no mass. At small distances, or equivalently high energies, the symmetry therefore appears to be valid. At large distances, however, the weak charge acts in some respects like a frictional force that would slow the particles down. Only at low energies, or equivalently large distances, does the Higgs field seem to give particles mass.

And this is exactly as we need it to be. The dangerous interactions that wouldn't make sense for massive particles apply only at high energies. At low energies particles can—and must, according to experiments—have mass. The Higgs mechanism, which spontaneously breaks the weak force symmetry, is the only way we know to accomplish this task.

Although we have not yet observed the particles responsible for the Higgs mechanism that is responsible for elementary particle masses, we do have experimental evidence that the Higgs mechanism applies in nature. It has already been seen many times in a completely different context—namely, in *superconducting* materials. Superconductivity occurs when electrons pair up and these pairs permeate a material. The so-called *condensate* in a superconductor consists of electron pairs that play the same role that the Higgs field does in our example above.

But rather than carry weak charge, the condensate in a superconductor carries electric charge. The condensate therefore gives mass to the photon that communicates electromagnetism inside the superconducting material. The mass *screens* the charge, which means that inside a superconductor, electric and magnetic fields do not reach very far. The force falls off very quickly over a short distance. Quantum mechanics and special relativity tell us that this *screening distance* inside a superconductor is the direct result of a photon mass that exists only inside the superconducting substrate. In these materials, electric fields can't penetrate farther than the screening distance because in bouncing off the electron pairs that permeate the superconductor, the photon acquires a mass.

The Higgs mechanism works in a similar fashion. But rather than electron pairs (carrying electric charge) permeating the substance, we predict there is a Higgs field (that carries weak charge) that permeates the vacuum. And instead of a photon acquiring mass that screens electric charge, we find the weak gauge bosons acquire mass that screens weak charge. Because weak gauge bosons have nonzero mass, the weak force is effective only over very short distances of subnuclear size.

Since this is the only consistent way to give gauge bosons masses, physicists are fairly confident that the Higgs mechanism applies in nature. And we expect that it is responsible not just for the gauge boson masses, but for the masses of all elementary particles. We know of no other consistent theory that permits the Standard Model weakly charged particles to have mass.

This was a difficult section with several abstract concepts. The notions of a Higgs mechanism and a Higgs field are intrinsically linked to quantum field theory and particle physics and are remote from phenomena we can readily visualize. So let me briefly summarize some of the salient points. Without the Higgs mechanism, we would have to forfeit sensible high-energy predictions or particle masses. Yet both of these are essential to the correct theory. The solution is that symmetry exists in the laws of nature, but can be spontaneously broken by the nonzero value of a Higgs field. The broken symmetry of the vacuum allows Standard Model particles to have nonzero masses. However, because spontaneous symmetry breaking is associated with an energy (and length) scale, its effects are relevant only at low energies—the energy scale of elementary particle masses and smaller (and the weak length scale and bigger). For these energies and masses, the influence of gravity is negligible and the Standard Model (with masses taken into account) correctly describes particle physics measurements. Yet because symmetry is still present in the laws of nature, it allows for sensible high-energy predictions. Plus, as a bonus, the Higgs mechanism explains the photon's zero mass as a result of its not interacting with the Higgs field spread throughout the universe.

However, successful as they are theoretically, we have yet to find experimental evidence that confirms these ideas. Even Peter Higgs has acknowledged the importance of such tests. In 2007, he said that he finds the mathematical structure very satisfying but "if it's not verified experimentally, well, it's just a game. It has to be put to the test."[60] Since we expect that Peter Higgs' proposal is indeed correct, we anticipate an exciting discovery within the next few years. The evidence should appear at the LHC in the form of a particle or particles, and, in the simplest implementation of the idea, the evidence would be the particle known as the *Higgs boson*.

THE SEARCH FOR EXPERIMENTAL EVIDENCE

"Higgs" refers to a person and to a mechanism, but to a putative particle as well. The Higgs boson is the key missing ingredient of the Standard Model.[61] It is the anticipated vestige of the Higgs mechanism that we expect that LHC experiments will find. Its discovery would confirm theoretical considerations and tell us that a Higgs field indeed permeates the vacuum. We have good reasons to believe the Higgs mechanism is at work in the universe, since no one knows how to construct a sensible theory with fundamental particle masses without it. We also believe that some evidence for it should soon appear at the energy scales the LHC is about to probe, and that evidence is likely to be the Higgs boson.

The relationship between the Higgs field, which is part of the Higgs mechanism, and the Higgs boson, which is an actual particle, is subtle—but is very similar to the relationship between an electromagnetic field and a photon. You can feel the effects of a classical magnetic field when you hold a magnet close to your refrigerator, even though no actual physical photons are being produced. A classical Higgs field—a field that exists even in the absence of quantum effects—spreads throughout space and can take a nonzero value that influences particle masses. But that nonzero value for the field can also exist even when space contains no actual particles.

However, if something were to "tickle" the field—that is, add a little energy—that energy could create fluctuations in the field that lead to particle production. In the case of an electromagnetic field, the particle that would be produced is the photon. In the case of the Higgs field, the particle is the Higgs boson. The Higgs field permeates space and is responsible for electroweak symmetry breaking. The Higgs particle, on the other hand, is created from a Higgs field where there is energy—such as at the LHC. The evidence that the Higgs field exists is simply that elementary particles have mass. The discovery of a Higgs boson at the LHC (or anywhere else it could be produced) would confirm our conviction that the Higgs mechanism is the origin of those masses.

Sometimes the press calls the Higgs boson the "God particle," as do many others who seem to find the name intriguing. Reporters like the term because people pay attention, which is why the physicist Leon Lederman was encouraged to use it in the first place. But the term is just a name. The Higgs boson would be a remarkable discovery, but not one whose moniker should be taken in vain.

Although it might sound overly theoretical, the logic for the existence of a new particle playing the role of the Higgs boson is very sound. In addition to the theoretical justification mentioned above, consistency of the theory with massive Standard Model particles requires it. Suppose only particles with mass were part of the underlying theory, but there was no Higgs mechanism to explain the mass. In that case, as the earlier part of the chapter explained, predictions for the interactions of high-energy particles would be nonsensical—and even suggest probabilities that are greater than one. Of course we don't believe that prediction. The Standard Model with no additional structures has to be incomplete. The introduction of additional particles and interactions is the only way out.

A theory with a Higgs boson elegantly avoids high-energy problems. Interactions with the Higgs boson not only change the prediction for high-energy interactions, they exactly cancel the bad high-energy behavior. It's not a coincidence, of course. It's precisely what the Higgs mechanism guarantees. We don't yet know for sure that we have correctly predicted the true implementation of the Higgs mechanism in nature, but physicists are fairly confident that some new particle or particles should appear at the weak scale.

Based on these considerations, we know that whatever saves the theory, be it new particles or interactions, cannot be overly heavy or happen at too high an energy. In the absence of additional particles, flawed predictions would already emerge at energies of about 1 TeV. So not only should the Higgs boson (or something that plays the same role) exist, but it should be light enough for the LHC to find. More precisely, it turns out that unless the Higgs boson is less than about 800 GeV, the Standard Model would make impossible predictions for high-energy interactions.

In reality, we expect the Higgs boson to be a good deal lighter than that. Current theories favor a relatively light Higgs boson—most theoretical clues point to a mass just barely in excess of the current mass bound from the LEP experiments of the 1990s, which is 114 GeV. That was the highest-mass Higgs boson LEP could possibly produce and detect, and many people thought they were on the verge of finding it. Most physicists today expect the Higgs boson mass to be very close to that value, and probably no heavier than about 140 GeV.

The strongest argument for this expectation of a light Higgs boson is based on experimental data—not simply searches for the Higgs boson itself, but measurements of other Standard Model quantities. Standard Model predictions accord with measurements spectacularly well, and even small deviations could affect this agreement. The Higgs boson contributes to Standard Model predictions through quantum effects. If it's overly heavy, those effects would be too large to get agreement between theoretical predictions and data.

Recall that quantum mechanics tells us that virtual particles contribute to any interaction. They briefly appear and disappear from whatever state you started with and contribute to the net interaction. So even though many Standard Model processes don't involve the Higgs boson at all, Higgs particle exchange influences all the Standard Model predictions, such as the rate of decay of a *Z* gauge boson to quarks and leptons and the ratio between the *W* and *Z* masses. The size of the Higgs's virtual effects on these *precision electroweak* tests depends on its mass. And it turns out the predictions work well only if the Higgs mass is not too big.

The second (and more speculative) reason to favor a light Higgs boson has to do with a theory called supersymmetry that we'll turn to shortly. Many physicists believe that supersymmetry exists in nature, and according to supersymmetry, the Higgs boson mass should be close to that of the measured *Z* gauge boson's and hence relatively light.

So given the expectation that the Higgs boson is not very heavy, you can reasonably ask why we have seen all the Standard

Model particles but we have not yet seen the Higgs boson. The answer lies in the Higgs boson's properties. Even if a particle is light, we won't see it unless colliders can make it and detect it. The ability to do so depends on its properties. After all, a particle that didn't interact at all would never be seen, no matter how light it was.

We know a lot about what the Higgs boson's interactions should be because the Higgs boson and Higgs field, though different entities, interact similarly with other elementary particles. So we know about the Higgs field's interactions with elementary particles from the size of their masses. Because the Higgs mechanism is responsible for elementary particle masses, we know the Higgs field interacts most strongly with the heaviest particles. Because the Higgs boson is created from the Higgs field, we know its interactions too. The Higgs boson—like the Higgs field—interacts more strongly with the Standard Model particles that have the biggest mass.

This greater interaction between a Higgs boson and heavier particles implies that the Higgs boson would be more readily produced if you could start off with heavy particles and collide them to produce a Higgs boson. Unfortunately for Higgs boson production, we don't start off with heavy particles at colliders. Think about how the LHC might make Higgs bosons—or any particles for that matter. LHC collisions involve light particles. Their small mass tells us that the interaction with the Higgs particle is so minuscule that if there were no other particles involved in Higgs production, the rate would be far too low to detect anything for any collider we have built so far.

Fortunately, quantum mechanics provides alternatives. Higgs

production proceeds in a subtle manner at particle colliders that involves virtual heavy particles. When light quarks collide together, they can make heavy particles that subsequently emit a Higgs boson. For example, light quarks can collide to produce a virtual W, the first picture in gauge boson. This virtual particle can then emit a Higgs boson. (See the first picture in Figure 51 for this production mode.) Because the *W* boson is so much heavier than either the up or down quarks inside the proton, its interaction with the Higgs boson is significantly greater. With enough proton collisions, the Higgs boson should be produced in this manner.

Another mode for Higgs production occurs when quarks emit two virtual weak gauge bosons, which then collide to produce a single Higgs, as seen in the second picture of Figure 51. In this case, the Higgs is produced along with two jets associated with the quarks that scatter off when the gauge bosons are emitted. Both this and the previous production mechanism produce a Higgs but also other particles. In the first case, the Higgs is produced in conjunction with a gauge boson. In the second case, which will be more important at the LHC, the Higgs boson is produced along with jets.

But Higgs bosons can also be made all by themselves. This happens when gluons collide together to make a top quark and an antitop quark that annihilate to produce a Higgs boson, as seen in the third picture. Really, the top quark and antiquark are virtual quarks that don't last a long time, but quantum mechanics tells us this process occurs reasonably often since the top quark interacts so strongly with the Higgs. This production mechanism, unlike the two we just discussed, leaves no trace aside from the Higgs particle, which then decays.

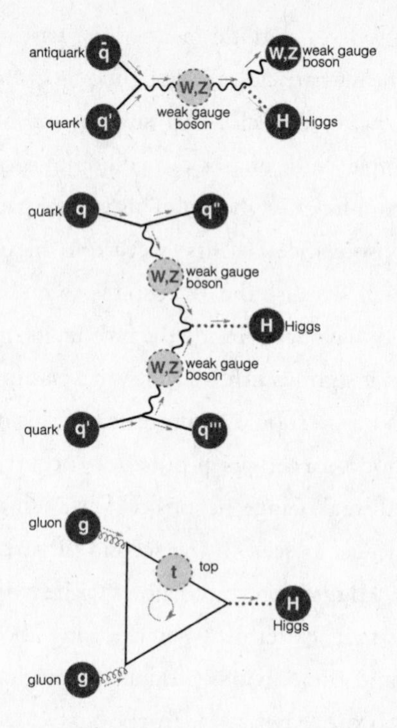

FIGURE 51. **Three modes of Higgs production: in order (*top to bottom*), Higgs-strahlung, *WZ* fusion, and *gg* fusion.**

So even though the Higgs itself is not necessarily very heavy—again, it is likely to have mass comparable to the weak gauge bosons and less than that of the top quark—heavy particles such as gauge bosons or top quarks are likely to be involved in its production. Higher-energy collisions, such as those at the LHC, therefore help facilitate Higgs boson production, as does the enormous rate of particle collisions.

But even with a big production rate, another challenge to observing the Higgs boson persists—the manner in which it decays. The Higgs boson, like many other heavier particles, is not stable. Note that it is a Higgs particle, and not the Higgs field, that decays. The Higgs field spreads throughout the vacuum to give mass to elementary particles and doesn't disappear. The Higgs boson is an actual particle. It is the detectable experimental consequence of the Higgs mechanism. Like other particles, it can be produced in colliders. And like other unstable particles, it doesn't last forever. Because the decay happens essentially immediately, the only way to find a Higgs boson is to find its decay products. The Higgs boson decays into the particles with which it interacts—namely, all the particles that acquire mass through the Higgs mechanism and that are sufficiently light to be produced. When a particle and its antiparticle emerge from Higgs boson decay, those particles must each weigh less than half its mass in order to conserve energy. The Higgs particle will decay primarily into the heaviest particles it can produce, given this requirement. The problem is that this means that relatively light Higgs boson only rarely decays into the particles that are easiest to identify and observe.

If the Higgs boson defies expectations and is not light, but turns out to be heavier than twice the *W* boson mass (but less than twice the top quark mass), the Higgs search will be relatively simple. The Higgs with a big enough mass would decay to the *W* bosons or *Z* bosons practically all the time. (See Figure 52 for decay into *W*s.) Experimenters know how to identify the *W*s and *Z*s that would remain, so Higgs discovery wouldn't be very hard.

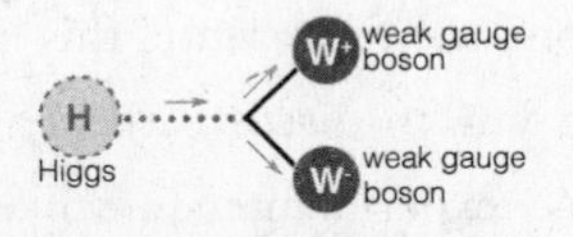

FIGURE 52. A heavy Higgs boson can decay to *W* gauge bosons.

The next most likely decay mode in this relatively heavy Higgs scenario would involve a bottom quark and its antiparticle. However, the rate for the decay into a bottom quark and its antiparticles would be much smaller because the bottom quark has much smaller mass—and hence much smaller interaction with the Higgs boson—than the *W* gauge boson. A Higgs heavy enough to decay into *W*s will turn into bottom quarks less than one percent of the time. Decays to lighter particles would happen less frequently still. So if the Higgs boson is relatively heavy—heavier than we expect—it will decay to weak gauge bosons. And those decays would be relatively easy to see.

However, as suggested earlier, theory coupled with experimental data about the Standard Model tell us the Higgs boson is likely to be so light that it won't decay into weak gauge bosons. The most frequent decay in this case would be into a bottom quark in conjunction with its antiparticle—the bottom antiquark (see Figure 53)—and this decay is challenging to observe. One problem is that when protons collide, lots of strongly interacting quarks and gluons are produced. And these can easily be confused with the small number of bottom quarks that will emerge from a hypothetical Higgs boson decay. On top of that, so many top quarks will be produced at the LHC that their decays to bottom quarks will also mask the Higgs signal. Theorists and experimenters are hard at work trying to see if there is any way to harness the bottom-antibottom final state of Higgs decay. Even so, despite the bigger rate, this mode probably isn't the most promising way to discover the Higgs at the LHC—though theorists and experimenters are likely to find ways to capitalize on it.

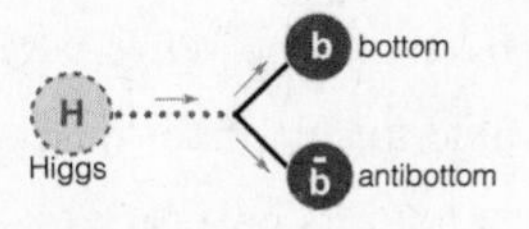

FIGURE 53. A light Higgs boson will decay primarily to bottom quarks.

So experimenters have to investigate alternative final states from Higgs decays, even though they will occur less frequently. The most promising candidates are tau-antitau or a pair of photons. Recall that taus are the heaviest of the three types of charged leptons and are the heaviest particles aside from bottom quarks that a Higgs boson can decay into. The rate to photons is much smaller—Higgs bosons decay into photons only through quantum virtual effects—but photons are relatively easy to detect. Although the mode is challenging, experiments will be able to measure photon properties so well once enough Higgs bosons decay that they will indeed be able to identify the Higgs boson that decays into them.

In fact, because of the criticality of Higgs discovery, CMS and ATLAS put elaborate and careful search strategies in place to find photons and taus, and the detectors in both experiments were constructed with a view to detecting the Higgs boson in mind. The electromagnetic calorimeters described in Chapter 13 were designed to carefully measure photons while the muon detectors help register decays of the even heavier taus. Together these modes are expected to establish the Higgs boson's existence, and once enough Higgs bosons are detected, we'll learn about its properties.

Both production and decay pose challenges for Higgs boson

discovery. But theorists and experimenters and the LHC itself should all be up to the challenge. Physicists hope that within a few years, we will be able to celebrate the discovery of the Higgs boson and learn more about its properties.

HIGGS SECTORS

So we expect to soon find the Higgs boson. In principle, it could be produced in the initial LHC run at half the intended energy, since that is more than sufficient to create the particle. However, we have seen that the Higgs boson will be produced from proton collisions only a small fraction of the time. This means that Higgs particles will be created only when there are many proton collisions—which means high luminosity. The original number of collisions that were scheduled before the LHC would shut down for a year and a half to prepare for its target energy was most likely too small to make enough Higgs bosons to see, but the plan for the LHC to run through 2012 before a year-long shutdown might permit access to the elusive Higgs boson. Certainly, when the LHC runs at full capacity, the luminosity will be high enough and the Higgs boson search will be one of its principal goals.

The search might seem superfluous if we are so confident that the Higgs boson exists (and if the pursuit is so difficult). But it's worth the effort for several reasons. Perhaps most significant, theoretical predictions take us only so far. Most people rightfully trust and believe only in scientific results that have been verified through observations. The Higgs boson is a very different particle from anything anyone has ever discovered. It would be the only fundamental scalar ever observed. Unlike particles such as quarks and gauge bosons, scalars—which are particles with zero spin—remain the same when you rotate or boost your system. The only spin-0 particles that have been observed so far are bound states of particles such as quarks that do have nonzero spin. We won't know for certain that a Higgs scalar exists until it emerges and leaves visible evidence in a detector.

Second, even if and when we find the Higgs boson and know for certain of its existence, we will want to know its properties. The mass is the most significant unknown. But learning about its decays is also important. We know what we expect, but we need to measure whether data agree with predictions. This will tell us whether our simple theory of a Higgs field is correct or whether it is part of a more complicated theory. By measuring the Higgs boson's properties, we will gain insights into what else might lie beyond the Standard Model.

For example, if there were two Higgs fields responsible for electroweak symmetry breaking rather than one, it could significantly alter the Higgs boson interactions that would be observed. In alternative models, the rate for Higgs boson production could be different than anticipated. And if other particles charged under Standard Model forces exist, they could influence the relative decay rates of the Higgs boson into the possible final states.

This brings us to the third reason to study the Higgs boson—we don't yet know what really implements the Higgs mechanism. The simplest model—the one this chapter has focused on so far—tells us that the experimental signal will be a single Higgs boson. However, even though we believe the Higgs mechanism is responsible for elementary particle masses, we aren't yet confident about the precise set of particles involved in implementing it. Most people still think we are likely to find a light Higgs boson. If we do, it will be an important confirmation of an important idea.

But alternative models involve more complicated Higgs sectors with an even richer set of predictions. For example, supersymmetric models—to be further considered in the following chapter—predict more particles in the Higgs sector. We would still expect to find the Higgs boson, but its interactions would differ from a model with only a single Higgs particle. On top of that, the other particles in the Higgs sector could give interesting signatures of their own if they are light enough to be produced.

Some models even suggest that a fundamental Higgs scalar does not exist but that the Higgs mechanism is implemented by a more complicated particle that is not fundamental but is rather a bound state of more elementary particles—akin to the paired electrons that give mass to the photon in a superconducting material. If this is the case, the bound state Higgs particle should be surprisingly heavy and have other interaction properties that distinguish it from a fundamental Higgs boson. These models are currently disfavored, since they are hard to match to all experimental observations. Nonetheless, LHC experimenters will search to make sure.

THE HIERARCHY PROBLEM OF PARTICLE PHYSICS

And the Higgs boson is only the tip of the iceberg for what the LHC might find. As interesting as a Higgs boson discovery will be, it is not the only target of LHC experimental searches. Perhaps the chief reason to study the weak scale is that no one thinks the Higgs boson is all that remains to be found. Physicists anticipate that the Higgs boson is but one element of a much richer model that could teach us more about the nature of matter and perhaps even space itself.

This is because the Higgs boson and nothing else leads to another enormous enigma known as the *hierarchy problem.* The hierarchy problem concerns the question of why particle masses—and the Higgs mass in particular—take the values that they do. The weak mass scale that determines elementary particle masses is ten thousand trillion times smaller than another mass scale—the Planck mass that determines the strength of gravitational interactions. (See Figure 54.)

The enormity of the Planck mass relative to the weak mass corresponds to the feebleness of gravity. Gravitational interactions depend on the *inverse* of the Planck mass. If it is as big as we know to be the case, gravity must be extremely weak.

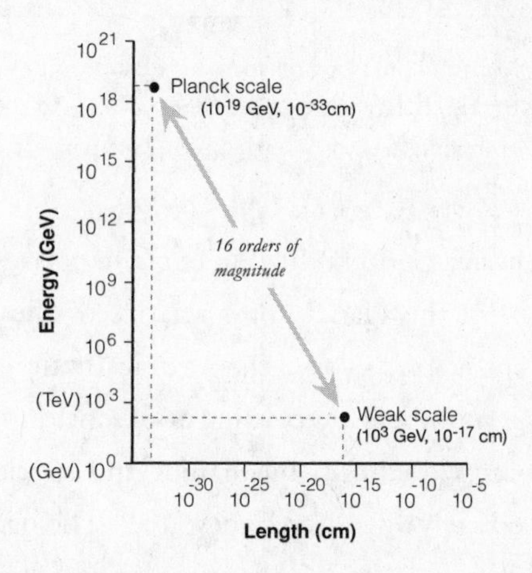

FIGURE 54. The hierarchy problem of particle physics: The weak energy scale is 16 orders of magnitude smaller than the Planck scale associated with gravity. The Planck length scale is correspondingly shorter than the distances probed by the LHC.

The fact is that fundamentally, gravity is by far the weakest known force. Gravity might not seem feeble, but that's because the entire mass of the Earth is pulling on you. If you were instead to consider the gravitational force between two electrons, you would find the force of electromagnetism is 43 orders of magnitude larger. That is, electromagnetism wins out by 10 million trillion trillion trillion. Gravity acting on elementary particles is completely negligible. The hierarchy problem in this way of thinking is: Why is gravity so much more feeble than the other elementary forces we know?

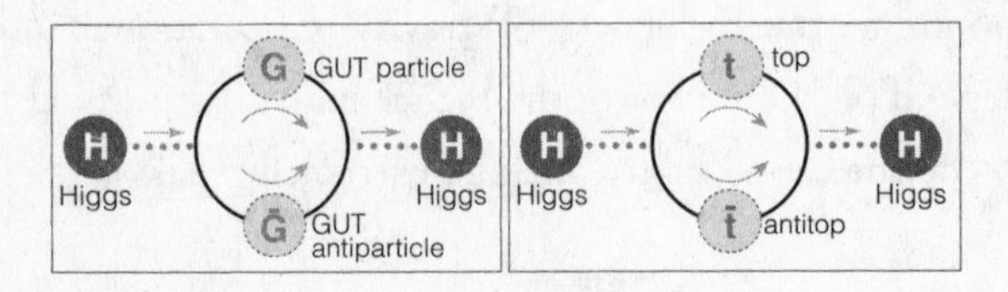

FIGURE 55. Quantum contribution to the Higgs boson mass from a heavy particle—for example with GUT-scale masses—and its antiparticle (*left*) and from a virtual top quark and its antiparticle (*right*).

Particle physicists don't like unexplained large numbers, such as the size of the Planck mass relative to the weak mass. But the problem is even worse than an aesthetic objection to mysterious large numbers. According to quantum field theory, which incorporates quantum mechanics and special relativity, there should be barely any discrepancy at all. The urgency of the hierarchy problem, at least for theorists, is best understood in these terms. Quantum field theory indicates that the weak mass and the Planck mass constant should be about the same.

In quantum field theory, the Planck mass is significant not

only because it is the scale at which gravity is strong. It is also the mass at which both gravity and quantum mechanics are essential and physics rules as we know them must break down. However, at lower energies, we do know how to do particle physics calculations using quantum field theory, which underlies many successful predictions that convince physicists that it is correct. In fact, the best measured numbers in all of science agree with predictions based on quantum field theory. Such agreement is no accident.

But the result when we apply similar principles to incorporate quantum mechanical contributions to the Higgs mass due to virtual particles is extraordinarily perplexing. The virtual contributions from just about any particle in the theory seem to give a Higgs particle a mass almost as big as the Planck mass. The intermediate particles could be heavy objects, such as particles with enormous GUT-scale masses (see left-hand side of Figure 55) or the particles could be ordinary Standard Model particles, such as top quarks (see right-hand side). Either way, the virtual corrections would make the Higgs mass much too large. The problem is that the allowed energies for the virtual particles being exchanged can be as big as the Planck energy. When this is true, the Higgs mass contribution too can be almost this large. In that case, the mass scale at which the symmetry associated with the weak interactions is spontaneously broken would also be the Planck energy, and that is 16 orders of magnitude—ten thousand trillion times—too high.

The hierarchy problem is a critically important issue for the Standard Model with only a Higgs boson. Technically, a loophole does exist. The Higgs mass, in the absence of virtual contributions, could be enormous and have exactly the value that would cancel the virtual contributions to just the level of precision we need. The problem is that—although possible in principle—this would mean 16 decimal places would have to be canceled. That would be quite a coincidence.

No physicist believes this fudge—or fine-tuning as we call it. We all think the hierarchy problem, as this discrepancy between masses is known, is an indication of something bigger and better in the underlying theory. No simple model seems to address the problem completely. The only promising answers we have involve extensions of the Standard Model with some remarkable features. Along with whatever implements the Higgs mechanism, the solution to the hierarchy problem is the chief search target for the LHC—and the subject of the following chapter.

NOTES

HIGGS DISCOVERY

1. This is a subtle point, but you might want to know that despite the Higgs mechanism's importance, it does not account for most of the mass in the universe. The preponderance of mass in ordinary matter (not dark matter) comes from the mass in the nuclei of atoms and is well understood. That mass arises because of the influence of the strong force, which provides energy when it binds together the three quarks inside protons and neutrons. Due to $E = mc^2$, that energy is equivalent to mass. However, particles that don't experience the strong force, such as the electron, would not have mass without the Higgs mechanism. Neither would the elementary quarks themselves nor the weak gauge bosons that communicate the weak force.
2. More precisely, the decay product is an antineutrino, the antiparticle of the associated neutrino.
3. The appended chapters discuss symmetry-breaking further.
4. Much more about the LHC and the CMS and ATLAS experiments can be found in *Knocking on Heaven's Door.*

EXCERPT FROM *WARPED PASSAGES*

17. To make a Higgs model work, at least one of the Higgs fields must be forced to take a nonzero value. This would be true if the minimum energy configuration occurs when the value of at least one of the Higgs fields is nonzero. One way this can happen is illustrated in Figure M2, which shows the so-called Mexican hat potential, a plot of

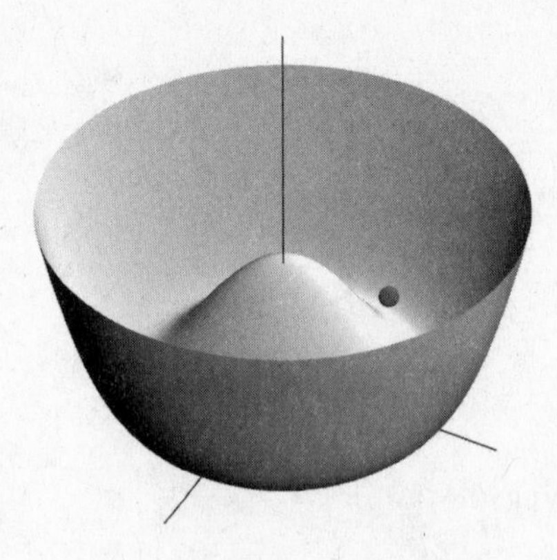

Figure M2. *The "Mexican hat" potential for the Higgs field.*

the energy the system would take for any combination of values of the two Higgs fields, where the two lower axes are the absolute values of the two Higgs fields and the height of the three-dimensional surface represents the energy of that particular configuration. This particular potential takes the form $\lambda(|H_1|^2 + |H_2|^2 - v^2)^2$, where λ determines how bowed up the potential is and v determines the value that $|H_1|^2 + |H_2|^2$ will take when the potential is at its minimum. The key feature of this potential is that when both fields have zero value, it is at a local maximum. Therefore, energy considerations tell us that the Higgs fields will not both be zero. Instead, they will take values that put them at the bottom of the circular basin surrounding the origin.

18. A more accurate way to describe the weak force symmetry would be to say that it rotates fields, rather than interchanging them.
19. This actually simplifies the symmetry breaking. Even if both x and y were nonzero—if both x and y were 5, for example—the rotational symmetry would be broken since a particular direction is picked out, the direction pointing from x = 0, y = 0 to the point where x = 5 and y = 5. A similar "rotational" symmetry applies to Higgs1 and Higgs2, but I have simplified and described the symmetry simply as an interchange symmetry. In the true description, even if both Higgs fields take the same value, the weak interaction symmetry would be

broken—in much the same way as the point x = 5, y = 5 spontaneously breaks rotational symmetry.

20. Although this model starts with two complex Higgs fields, there is only a single Higgs particle in the end. That is because the three other (real) fields become the three additional fields that are required to turn three massless particles with two physical polarizations into massive particles with three polarizations. Three of the Higgs fields become the third polarizations of the three heavy weak gauge bosons—the two Ws and the Z. The fourth remaining Higgs field should create true physical Higgs particles. If this model is right, the LHC should produce them.

EXCERPT FROM *KNOCKING ON HEAVEN'S DOOR*

60. On WNYC's *The Takeaway,* March 31, 2007.
61. Sometimes people also debate whether right-handed neutrinos belong in the Standard Model. Even if present, they are likely to be extremely heavy and not very important for lower-energy processes.

ACKNOWLEDGMENTS

This Ecco Solo happened very quickly. I'd like to thank Bob Cahn, David Krohn, Andi Machl, Luboš Motl, and Matthew Reece for their rapid read-throughs and valuable comments. I'd also like to thank Frank Close, Melissa Franklin, Jean-Marie Frère, Fabiola Gianotti, Rolf Heuer, Joe Incandela, Joe Lykken, and Steve Myers for answering my questions. Thanks to my editors, Hilary Redmon and Daniel Halpern, for introducing the Solo idea and their helpful edits and follow-through, and also to Shanna Milkey and the others at Ecco who contributed. Thanks also to my agents, Andrew Wylie, Sarah Chalfant, and James Pullen, for solidifying the project. And of course thanks to all the amazing engineers and physicists who continue to make knowledge advance—and to all those who stayed up late at night to get this result out as quickly as they did.